GRANDMA CALLED IT
Roughage

GRANDMA

CALLED IT *Roughage*

Fiber Facts and Fallacies

Robert L. Ory

American Chemical Society, Washington, D.C., 1991

Library of Congress Cataloging-in-Publication Data

Ory, Robert L., 1925–
Grandma called it roughage: fiber facts and fallacies/Robert L. Ory p. cm.

Includes bibliographical references (p.) and index.

ISBN 0–8412–1749–1 Hardcover.—ISBN 0–8412–1764–5 (pbk).
1. Fiber in human nutrition. I. Title.

TX553.F53079 1991
613.2′.8—dc20

91—24259
CIP

The paper used in this publication meets the minimum requirements of American National Standard for Information Sciences—Permanence of Paper for Printed Library Materials, ANSI Z39.48—1984.

PRINTED IN THE UNITED STATES OF AMERICA

To Anne

for many years of love,

patience, and understanding

Contents

Preface *ix*

1
Fibermania and How It Began *3*

2
What Is Fiber and Where Do We Find It? *11*

3
Fiber and Metabolism vs. Fat and Cholesterol *25*

4
Can Fiber Fight Fat? *35*

5
Fiber and the Big "C"—Constipation *47*

6
Can Fiber Cure Cancer, Diabetes, and Other Serious Diseases? *61*

7

Fiber and Mineral Nutrition *75*

8

Is Fiber an Essential Nutrient? *89*

9

Breakfast Cereals: The "Battle of the Brans" *99*

10

Fiber in Fruits and Vegetables *111*

11

Can Fiber Be Good for You and Still Taste Good? *123*

Bibliography *149*

Index *155*

Preface

Many technical books about dietary fiber have been written by qualified scientists during the past 10 to 15 years, and fiber has received a good deal of publicity in newspapers, magazines, and other nontechnical publications, as well. Nevertheless, so many consumers are misinformed and willing to believe not-quite-accurate statements about dietary fiber and the miraculous cures attributed to it, that I felt that a semitechnical but simply written book on what fiber is, what it does, and what it does not do was needed to dispel fears and to correct some misconceptions. So, after 15 years of chemical and biochemical research on dietary fiber—primarily rice and rice bran, but also wheat bran, peanuts, and several vegetables—I decided to write an easy-to-read book on what we know about the benefits of fiber on the big "C" words: constipation, cholesterol, cancer, and colonic and coronary diseases. I've tried to write it in an understandable and interesting style for easy reading by both scientists and nonscientists. In order to support or elucidate upon statements and findings about fiber's properties, I've included several tables of data and figures drawn from research on dietary fiber done by scientists and reported in technical journals and books.

Much of the stimulus for writing this book evolved from my coming home from the laboratory at night, excited and wanting to explain the day's research to my wife, a former elementary school teacher who hated chemistry. As each chapter was completed, it was given to her to read. If she, as a lay person, understood it and found it to be interesting, I would then go on to the next chapter.

This book is *not* intended to be the final word on dietary fiber. Research goes on continually, as it must if we are to improve our knowledge of proper nutrition for better health. This book *is* intended to quell rumors about dietary fiber and to inform those who don't understand enough about fiber's benefits and limitations. If I have accomplished those goals, then the two years I devoted to the literature search and the writing of this book will have been worth it!

ROBERT L. ORY
New Orleans, LA

June 1991

GRANDMA CALLED IT *Roughage*

1

Fibermania and How It Began

Since the early 1970s, Americans, and more recently Europeans, have jumped on the fiber bandwagon. But many people don't even know what it is, or what foods contain fiber. The term dietary fiber (or food fiber) means different things to different people. To some, it is the coarse indigestible bran coating of wheat. To others, it is the water-holding components of fruits, vegetables, and certain pharmaceutical bulking agents. To some, it is roughage, the stuff you can't digest but is good for your bowels. To scientists engaged in fiber research, it's all of these, and maybe a little bit more.

When I was a small boy more than 50 years ago, my grandmother (who lived in the country and never had a single course in nutrition) always prepared balanced meals that had to include a protein, a starch, a green and a yellow vegetable, and roughage. With today's technology and the ongoing research on foods and nutrition, we have learned a great deal about the chemical composition and functions of each of these dietary components. In general, only plant foods like cereals, vegetables, fruits, and nuts contain dietary fiber. Animal foods like milk, meat, fish, eggs, and poultry do not.

BRIEF HISTORY

The milling and refining of wheat grains to remove bran and produce a white milled flour for baking actually goes back to biblical times. But the virtues of bran were well-known then too. The father of modern medicine, Hippocrates, who gave his name to the physician's oath, and who lived almost to age 80 before dying in 377 B.C., recommended eating "unbolted wheat meal" bread (unmilled wheat flour containing the bran) for its effect on the bowels.

In the early 1800s a common complaint of Americans who ate too much fat and sugar was indigestion. Sylvester Graham, an American food reformist, claimed this complaint was the result of a too-concentrated diet. He recommended putting the bran back into the wheat flour as a means of correcting this problem. Graham's name is memorialized today in the graham cracker and in graham (that is, whole wheat) flour.

BURKITT'S WORK

Current interest in dietary fiber is due largely to the work of Dr. Denis Burkitt, a British physician who was the first to realize that what was once considered to be several different cancers in Africa was actually just one: a cancer of the lymphoid tissue, which became known as Burkitt's lymphoma. It is common where malaria is widespread and intense.

While practicing and doing research in Africa for 20 years, Dr. Burkitt became convinced that the African diet of coarse grains, cereals, yams, and beans was the primary reason for the low incidence of colon–rectal cancer there. He conducted research on the incidence of colon cancer in African bush populations in the 1960s–1970s, compared to populations living in London. He and his co-workers, Drs. Hugh Trowell and Kenneth Heaton, showed a connection between diet and some diseases common to Westerners. In Africa, where the diet was low in animal products but high in cereal grains and plant foods, disorders such as colon cancer, appendicitis, varicose veins, diverticulosis, and hemorrhoids were virtually absent. In Western cultures that consumed fewer cereals and fresh vegetables and more animal products, the reverse was true.

He also compared the amounts and consistence of fecal discharges of both societies. Africans excreted larger volumes of softer, bulkier stools than did Britishers on low-fiber diets, who excreted hard, pel-letlike stools. This research confirmed a major reason for ample amounts of fiber in the diet: Fiber increased water content of the feces and reduced or eliminated constipation, a chronic disease of Western populations.

Dr. Burkitt also noted that the Africans were generally thin compared to Westerners, and he concluded that fiber and the more energetic lifestyle of the African natives were probably the means of controlling obesity.

Since the publication of studies in various professional journals, fiber has jumped to *number one* on the dietary hit parade of Americans as a means of controlling diseases of the gastrointestinal (GI) tract, improving bowel regularity, and controlling weight.

Though many scientists today are engaged in research on all kinds of food fiber in all parts of the world, Dr. Burkitt is possibly the most vocal advocate of high-fiber–low-fat diets as a means of improving health and preventing most diseases of the GI tract. He said, "If people

went in more for whole meal bread and potatoes, they would drive the laxative firms out of business and cut hospital stays in half." On the basis of his years of research studying Africans living in rather primitive conditions and those who moved to the West and adopted Western

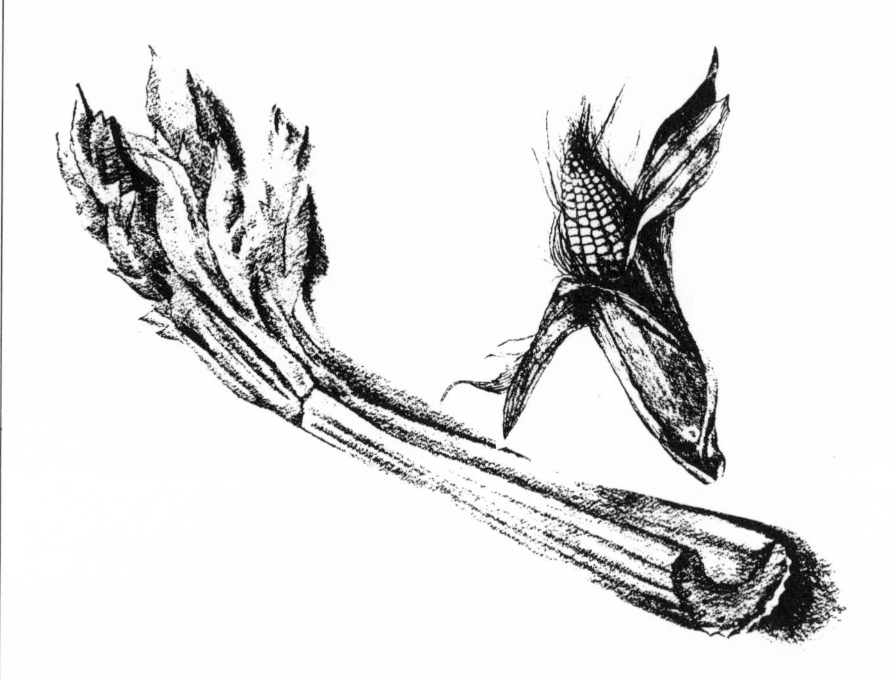

habits of living and eating, he proposed a theory: that eating foods high in fiber (unrefined whole grains, cereal brans, fresh fruits, and vegetables) and low in fat will prevent many digestive diseases such as colon cancer, hemorrhoids, diverticular disease, hiatal hernia, varicose veins, appendicitis, atherosclerosis, and certain circulatory disorders. These diseases are virtually unknown in bush- and rural-living Africans who consume coarsely ground grains, beans, peas, and little or no meat.

The birth of the "fiber" hypothesis in the late 1960s by Dr. Burkitt and his associates, Trowell and Walker, stimulated a huge increase in research on the chemistry and clinical effects of dietary fiber that continues today. Some of the prominent scientists who have pioneered in this area of research are D. Kritchevsky, G. Spiller, P. Van Soest, D. Southgate, M. Eastwood, R. Kay, and J. Kelsay, to name just a few.

Today Americans and Europeans are bombarded with high-powered advertisements by health-food advocates, food fadists, food product manufacturers, pharmaceutical manufacturers, and even lesser-known charlatans who offer us what we want to hear about fiber cure-alls at prices we're willing to pay.

WHAT IS FIBER?

What, then, is fiber and where do we get it? In plain terms, fiber is that portion of plant food that is not broken down by enzymes in the GI tract. It is those components of plant foods that pass through the digestive tract into the large intestine undigested or only partly digested. This undigestible portion of plant foods was called roughage by Grandma, and her term may be more descriptive than the term fiber. Fruit pectins, vegetable gums, and mucilages were certainly not fibrous, but they are food fibers. Dietary fiber does *not* include fibrous cuts of meat, or the tendons, ligaments, and bits of connective tissue in red meats. These foods are protein, and protein-splitting enzymes in the body can break them down to smaller units.

Proteins, fats, and starches are broken down in the digestive tract by enzymes (proteinases, peptidases, lipase, and amylase) that come from the stomach and the pancreas. Indigestible fiber passes through

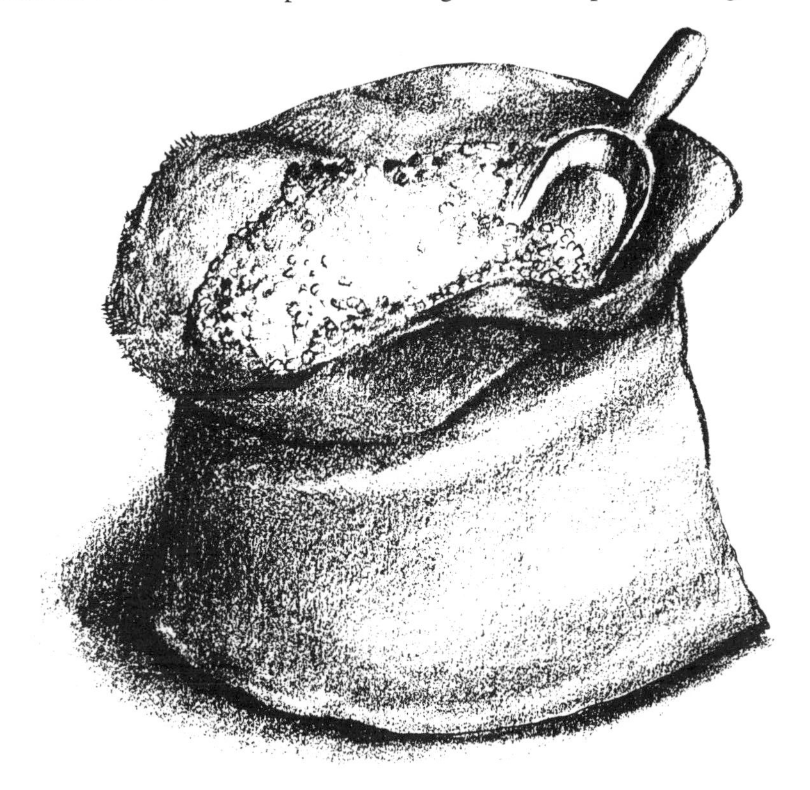

the stomach and small intestine essentially unchanged, but in the large intestine some forms of fiber may be partially broken down, as we will see later. The major types of dietary fiber are

- cellulose, hemicellulose (or pentosans), and lignin derived mainly from cereal grains, nuts, and oilseeds;
- pectins, gums, and mucilages derived mostly from fruits and vegetables.

The amounts and kinds of each type of fiber vary from plant to plant and between different plants of the same variety. Variations within a certain plant species depend upon the age of the plant, the conditions under which it was grown, how long it was stored after harvest, and the

"If you're looking for roughage in your food, you won't find it rougher anywhere!"

conditions of storage—temperature, product moisture content, and humidity. Thus, when speaking of fiber it is necessary to define the type (cellulose, lignin, or pectin,) and the source (the specific cereal grain, fruit, vegetable, or nut) in order to discuss its properties and health benefits.

As succeeding chapters will show, all fibers are not alike in their reactions and health benefits. Fiber from cereals is different from fruit and vegetable fibers. The composition and properties of wheat bran are completely different from those of apple pectin. Some fibers have excellent water-holding capacity, and so increase movement of food through the bowel. Some bind to bile acids and affect fat digestion and its subsequent resorption into the blood, or affect cholesterol levels in blood. Others are partially fermented in the large intestine. Some, like lignin, pass through the intestines completely unchanged.

The biological and physiological importance of dietary fiber went largely overlooked until the middle of the 20th century when Western cultures began to question the advantages and disadvantages of milling and refining the cereal grains, fruits, and vegetables that we consume. Research on dietary fiber has been widespread. We know much more than we did 25 or 30 years ago, but there are still misconceptions, misinformation, and some mystique in discussions of dietary fiber. The following chapters will help clarify some of these misconceptions about what fiber is and what it is not, what it does, and what it does not do.

2

What Is Fiber and Where Do We Find It?

The confusion in our understanding of what fiber is and what it does for human nutrition and health is not surprising when we see the complexity of this food component. Fiber is a part of all plant foods: cereal grains, fruits, vegetables, nuts, and oilseeds. It is found in the walls of the individual cells of every plant part and makes up the skeletal backbone structure of the roots, stalks, stems, leaves, flowers, fruits and seeds.

Except for lignin, which is not a carbohydrate, the chemical forms of fiber—cellulose, hemicellulose, pectins, gums, and mucilages—are nonstarch carbohydrates made up of thousands of sugar molecules joined to each other as huge polymers. A *polymer* is a large molecule composed of a large number of similar small molecules, called monomers, joined together by the same type of linkage (chemical bond). In general, the monomers can be joined in a straight chain, much like a long string of pearls, or in a branched chain, which resembles a tree branch. In fiber, the monomers are sugars, and the polymers are carbohydrates, also called polysaccharides.

For many years, analyses of foods for total composition, which described content in terms of protein, fat, moisture, ash or minerals, and carbohydrate, also contained a value for "crude fiber". That term is really a misnomer; crude fiber contains only lignin and part of the cellulose. This value did not include hemicelluloses, pectins, gums and mucilages. Thus, published values for fiber were low and not a true measure of the total amount of material—roughage—not digested by humans.

Crude fiber is still included in standard analyses of food components, but the term "dietary fiber" was introduced in the mid-1970s to reflect a more comprehensive analysis encompassing all of the non-digestible residue of plant foods that pass through the GI tract and into the colon. In this chapter, I will try (1) to describe the chemical make-up and the differences among various forms of fiber, and (2) to briefly show how food chemists have improved the analytical methods used to measure total dietary fiber in food samples today, not just the crude fiber.

CELLULOSE

Cellulose, the major component of plant cell walls, is perhaps the most well-known form of the various fibers. In fact, the long threadlike structure of cellulose probably gave rise to the term "fiber". Chemically, it consists entirely of straight chains of up to 10,000 glucose molecules joined together into large, insoluble polymers. Glucose is a simple six-carbon sugar easily metabolized by the body. But in cellulose the number of molecules joined together is so large that we can't digest the resulting polymer. Like glucose, cellulose consists of carbon, hydrogen, and oxygen only; it contains no nitrogen or sulfur. Some starches are chemically similar to cellulose in that they contain polymers of thousands of glucose molecules, but in a much different structure with very different properties. Starches are soluble in water, and they can be digested by amylase, the enzyme derived from the human pancreas.

Cows, sheep, and goats can easily utilize cellulose as a form of energy because of the microbes that live in their rumens, but animals with a single stomach, like pigs, chickens, and, of course, humans, do not have either cellulose-splitting enzymes in their bodies or the microbes contained in the ruminant stomachs.

Cellulose is the most prevalent carbohydrate in the plant kingdom and, in some plants, can be up to half the weight of the plant. Wood pulp and paper pulp, for example, consist mostly of cellulose; cotton taken directly from the bolls is almost pure cellulose. Once washed and cleaned to an essentially pure compound that is useful as a standard, called alpha-cellulose (Alphacel), cellulose in this form is sometimes used as a food ingredient to improve texture, creaminess or other physical properties of certain foods. It has been added to breads to raise the fiber content and nutrient density while lowering caloric content.

Today the food and baking industries use other food-grade derivatives of cellulose as much as or more than alpha-cellulose in various products. Carboxymethylcellulose and methylcellulose are water-soluble derivatives of cellulose used as thickening agents because of their high affinity for water. Because of this property, the cellulose derivatives behave more like gums than like the parent cellulose. However, W. B. Miller reported in 1986 that the powdered cellulose used by the food industry as a food ingredient is only about 90% cellulose; the other 10% is hemicellulose because the purification process for producing cellulose is nonselective.

The raw material for producing most food-grade cellulose is wood. Wood is an economical source, but any other plant source, such as bran or celery, could be used. In general, the trees are cut into logs, debarked, then cut into small chips that are subsequently cooked in large digesters at high temperatures and pressures in chemicals such as sodium bisulfide and sodium hydroxide. After cooking, the pulp is washed extensively, bleached in a series of chlorinations, extracted to remove any lignin residue, washed again, formed into a continuous sheet that is dried by heat, then formed into rolls or bales. Powdered cellulose is produced from these rolls or bales by chopping them into small pieces that are fed into ball mills for grinding to powders of various particle sizes, depending upon the needs of food manufacturers. The resulting white, flavorless powder is a desirable food ingredient that is insoluble in water but can hold large amounts of water—up to seven times its weight. It does not add undesirable flavors to any product, does not mask desirable flavors, and, because it is chemically inert, does not interfere with any reactions taking place in the food system. The fine grades of cellulose are best suited for baking. Coarser grades are more useful in fabricated foods because they help maintain the structure of the product. Powdered cellulose is on the U.S. Food and Drug Administration's list of food ingredients that are generally recognized as safe (GRAS). One of its fastest-growing uses is to lower calories and raise nutrient density in low-calorie food products, especially breads.

Powdered cellulose has also been added to cakes, cookies, doughnuts, and pie and pizza crusts as an ingredient to give bulk and improve texture and structure. It also improves texture and lowers calories in cake fillings, icing, confections, barbecue sauces, and salad dressings. Some diet drinks and nutritional supplements contain powdered cellulose to absorb moisture and provide texture in the reconstituted drinks. Puddings, frozen desserts, and toppings also include powdered cellulose to

maintain a thick creamy texture after thawing and to prevent formation of ice crystals while freezing. As new food products enter the market, the applications of powdered cellulose as a food ingredient will also grow.

HEMICELLULOSES

The term hemicellulose may be a misnomer. It originated at the turn of the century and was used to describe large chains of sugars that were extracted with dilute alkali from plant cell wall material. Today we know that they are not related to cellulose at all, but the term has stuck. Hemicelluloses, with pectin, form the matrix of the plant cell wall in which cellulose is intermingled. In terms of human health, their water-holding capacity, their partial digestibility in the colon, and their ability to bind minerals and bile acids makes hemicelluloses an important form of dietary fiber.

Hemicelluloses, like cellulose, are large structural polymers associated with plant cell walls, but they are less fibrous than cellulose and are composed of several other sugars besides glucose, plus varying numbers of amino acids, the building blocks of protein. Hemicelluloses contain mixtures of 50–200 five-carbon sugars (xylose and arabinose) and six-carbon sugars (glucose, galactose, mannose, and rhamnose), plus lesser amounts of the sugar acids (glucuronic and galacturonic acids). Attached to these sugars are varying amounts of amino acids in different-sized chains that add a proteinlike portion to the hemicelluloses. Hemicelluloses from cereal grains have also been referred to as pentosans when then contain large numbers of xylose and arabinose molecules, as they do in wheat.

During the 12 years that we conducted extensive research on the hemicelluloses of rice at the U.S. Department of Agriculture's Southern Regional Research Center in New Orleans, LA, we learned that the hemicelluloses are just as diverse as other kinds of food fibers. There are water-soluble and alkali-soluble (water-insoluble) forms that differ in chemical composition between the bran and the inner endosperm (white milled rice grain) layers, between the long-, medium-, and short-grain

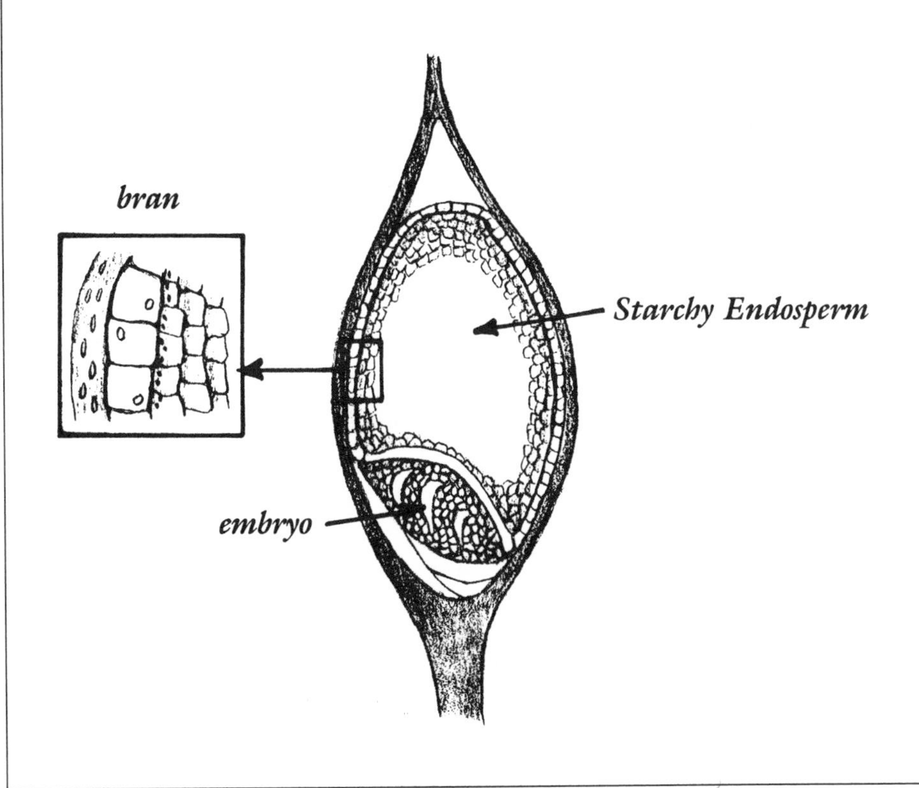

varieties, and between the same varieties grown in Texas, Louisiana and Arkansas. As we will see, these different structures and compositions affect the binding of the hemicelluloses to minerals and bile acids, so that some sources of hemicellulose appear to be more beneficial physiologically then others.

LIGNIN

Lignin is the nonpolysaccharide (not a polymer of sugars) among the various fibers, and is by far the most indigestible. The name "lignin" is derived from the Latin name for wood, *lignum*. Chemically, lignin consists of hundreds of repeating units of phenylpropane molecules in which phenyl, a cyclic six-carbon ring structure, is bound to the three-carbon propane chain. As a plant grows and matures, it produces increasing amounts of lignin, which acts as a cement to hold together other materials in the cell wall. This chemical makeup and hard structure protects lignin from biochemical degradation and digestion. With the exception of wheat bran products, the amount of lignin generally consumed by humans is rather low because preparation will usually remove or discard the tough or woody portions of plant foods for reasons of esthetics or palatability. Thus, lignin is not present in quantities as great as cellulose, pectins, hemicelluloses, gums, and mucilages.

However, lignin may have useful properties other than as a fiber. Catignani and Carter (1982) found that inclusions of 1–10% lignin in rat diets exhibited antioxidant properties in the presence of oil- and fat-soluble vitamins (A and E). Antioxidants are important because they inactivate oxygen-containing free radicals, which are thought to attack DNA (promoting cancer), artery walls (leading to heart disease), and other body parts (accelerating aging).

PECTIN

Pectin is a collective term given to a group of amorphous polysaccharides universally present in plants as part of the primary cell wall and the intercellular cement layers. Chemically, the pectins are composed primarily of hundreds of molecules of galacturonic acid, but they sometimes have other sugars bound to the galacturonic acid at various points in the large polymer. The addition of methyl groups ($-CH_3$) to some of the carboxyl groups ($-COOH$) of the galacturonic acid molecules in the large polymers confers different but desirable properties on the various pectins. They are generally water-soluble and hold large quantities of water.

The presence and the location of small amounts of sugars in the pectins can have significant effects on their gelling properties. The amounts present can very greatly, from small or even trace amounts in cereal brans to as much as 15–30% of dry weight in fruits and vegetables, such as apples or citrus. About 30% of dry weight of the white layer of citrus peel and 15–17% of dried apple pomace is pectin. Most of the commercial pectin produced today is derived from these two sources.

Unlike cellulose, pectins do not exist as threadlike fibers. Structurally more like soft gels or jellies (depending upon their water content), they are regarded as a form of dietary fiber because they pass through the GI tract undigested. As we shall see, their binding properties make them an excellent form of fiber, with some important benefits for good health. In general the fruit pectins are more soluble in hot water, but they do bind large amounts of cold water, especially in the presence of added sugar, as in jellies and jams. Pectins are present in smaller amounts than the other cell wall constituents (cellulose, hemicellulose, and lignin) and make up 1–4% of the total cell wall polysaccharides, i.e., that pectins are only 1–4% of the total fiber components, with little or no water. In general, they are most abundant in fruits and vegetables and least in cereal grains and nuts.

The old expression, "an apple a day keeps the doctor away", can probably be attributed to the high pectin content of apples. Fresh apples contain 0.5–1% pectin by weight, and in fresh apples the weight is almost all water. As a group, pectins are biochemically less well defined than the other forms of fiber, but recently several research reports have appeared. They describe the chemical composition, binding properties, and various reactions of pectin that has been isolated from specific sources. In addition to citrus and apples, pectins have been isolated from the seeds of sunflower, mustard, and lupin, and also from sugar

beets, onions, and soybeans. Pectin's ability to dissolve in hot water and form a gel upon cooling has attracted a lot of research attention. Though fruits and vegetables contain enzymes that can remove or break down pectin and produce soft, mushy spots on the bruised area of the plant, humans do not possess these enzymes, so pectins pass through the GI tract intact.

Another property of pectin, one which has affected the frozen juice industry, is its ability to lower the freezing temperature of juice concentrates. This property could affect flavor quality and storage stability of frozen juice concentrates kept in home freezers that are not set low enough below the freezing temperature (–4 to 0 °C or 26 to 32 °F).

GUMS AND MUCILAGES

Because these two forms of fiber are closely related, we'll consider them together. They are very heterogeneous, complex polysaccharides of the plant kingdom, not part of the cell wall matrix like pectins but generally nondigestible by humans. Chemically, gums are composed of large polymers of galactose, rhamnose, and arabinose sugar molecules with various types of side chains attached to the backbone. One of the most popular food gums, gum arabic, has a backbone of galactose sugar units with various side chains of galactose, rhamnose, and arabinose sugars. Variations in the side chains are usually specific for different species. Gum ghatti, another food gum additive, has alternating glucuronic acid and mannose units as the backbone, and has galactose and arabinose in the side chains.

Gum arabic and gum ghatti illustrate the diversity of the food gums, related to the varied sources from which they are obtained. From tree exudates we get gum arabic and gum tragacanth; from plant extracts come agar, carrageenan, and alginates; from beans we get guar gum and locust bean gum.

The most popular synthetic gum used as a food ingredient is xanthan gum, which is prepared by pure culture fermentation of a plant microorganism, *Xanthomonas campestris*, which produces the gum during fermentation. Xanthan gum consists of a backbone of glucoses similar to cellulose, but on alternate glucoses there are side chains consisting of one glucuronic acid and two mannose sugar units.

So many people today are afraid of "added chemicals" in their food, but they do not have sufficient knowledge about what chemicals are, which ones are bad, and which are good. People need a better understanding of "good" chemicals, such as soluble gums added as food ingredients. These are not harmful at the levels added, and they improve the food's texture and other properties.

In high doses, locust bean, guar, and arabic gums have been shown to lower blood serum cholesterol levels. But they are used at low levels in food products, and health claims for the food gums are not emphasized in advertising to avoid their being classified by the Food and Drug Administration as drugs rather than as food ingredients. For example, in 1989–1990, two major breakfast cereal manufacturers began promoting the fact that their brands of ready-to-eat cereals contain psyllium, a type of mucilage extracted from *Psyllium* seed husks. Psyllium has been used as a "natural-fiber laxative" and is the active ingredient in a well-known commercial laxative. Because the drug companies that produced the laxative had to obtain FDA approval to use psyllium as a "drug" product, they took the cereal producers to court to prevent their touting psyllium's health benefits as a food additive. Apparently, to the FDA, drugs cannot promote health, but food additives can. It is ironic that food gums can serve as additional forms of dietary fiber in food products if added in higher doses, but cannot be advertised as such or they can be labeled as drugs.

Consumers today accept fortification of foods with vitamins and minerals. Some even expect to have these chemicals added to foods like breakfast cereals. Yet they have a negative image of chemicals in general.

With the growing interest in dietary fiber, the food industry should try to educate consumers about the food gum additives, emphasizing that they are safe additives classified as "good" chemicals, similar to vitamins and minerals, not harmful drugs.

Vegetable gums usually have soluble dietary fiber contents of 75–80%, but some products have been produced with 80–95% soluble dietary fiber. Food gums are used in gravies, canned soups and sauces, salad dressings, frozen foods, processed meats, ice creams, sherberts, puddings, instant desserts and soup mixes, whipped creams and toppings, and baked goods, but in levels too low to be considered a source of soluble fiber.

Mucilages are similar to gums in that they are large polysaccharides extractable with water, and they have excellent water-holding capacity. There are two forms: neutral and acidic. Neutral forms include galactomannans in guar and alfalfa seeds, which are closely related in structure to the hemicellulose galactomannans and glucomannans. They contain glucose and mannose in 1:3 ratios and arabinoxylans (found in extracts of cereal grains and linseed) that contain arabinose and xylose. Studies of the acidic forms of mucilages are more limited. Two have been reported in linseed and in slippery elm bark. Because they contain the same sugars as the plant gums and have similar solubilities and water-holding properties, mucilages are frequently discussed along with the gums. They tend to form very viscous and sticky gels when mixed with water in foods. One example is okra gumbo, a Creole dish based upon okra, a popular Louisiana vegetable used to thicken the souplike dish.

CHEMICAL ANALYSIS OF FIBER IN FOOD

These descriptions of the various forms of food fiber illustrate the complexity of analyzing the total amount of nondigestible plant residue in our diet. Some are truly fibrous, some are amorphous, some are jellylike. None come from red meats or fibrous animal products, which are included in the diet for their high-quality protein, vitamins, and minerals such as iron and zinc. Crude fiber had been used in nutrient composition tables for years, but it is no longer sufficient because it misses the soluble forms of fiber. To better analyze total fiber in foods, several approaches are now used in place of, or in addition to, crude fiber analysis.

The crude fiber (CF) method involves extraction of the sample with acid and alkali after solvent extraction to remove fat. This method measures lignin and most of the cellulose. In the acid detergent fiber (ADF) method, approved in 1975, the sample is boiled in a detergent solution containing sulfuric acid. The ADF method also measures lignin and cellulose, but yields higher values for both than the CF method. Because of the difficulties incurred with samples containing high amounts of starch like cereal grains and potatoes, the neutral detergent fiber (NDF) method was approved in 1982. In this method, the enzyme alpha-amylase is added first to hydrolyze starch down to sugars that are easily removed by dialysis. The acid and alkali steps of CF are replaced with milder extraction steps that yield very good recoveries of the cellulose, lignin, and hemicelluloses. The NDF method is better for cereal grains and brans because these components account for essentially all of their dietary fiber. For fruits and vegetables, however, the NDF method may miss the water-soluble pectins, gums, and mucilages; it will pick up only larger, insoluble pectins that are sometimes present.

In 1981 several of the world's leading fiber scientists from 32 different laboratories in the United Stated and Europe collaborated to develop the total dietary fiber (TDF) method, which was accepted and approved in 1984. In the TDF method, duplicate samples are treated with multiple enzymes (alpha-amylase and amyloglucosidase to digest the starch and a protease to digest the protein) to remove starch and protein by mild methods. Then 78% ethyl alcohol is added to precipitate the soluble dietary fibers along with the insoluble forms already there. The total precipitated dietary fibers are then carefully washed with ethyl alcohol and acetone, dried, and weighed. This material is analyzed for any residual protein and ash (minerals), and these values are subtracted from the dried weight to obtain the total dietary fiber. This value includes the lignin, cellulose, hemicelluloses, pectins, gums, mucilages, and any modified celluloses or synthetic gums that may have been added to the food product as a food ingredient.

We have come a long way in the development of better methods to accurately estimate fiber contents of foods. But as Dr. Alfred Olson wrote in 1987,

> . . . rigorous comparisons between the methods have not been made. Research is needed to provide reference procedures that adequately quantitate and characterize soluble and insoluble dietary fiber. Also, new procedures may have to be developed to deal with different types of food, such as high-starch and low-starch products. Any analysis for dietary fiber must be related to what we understand dietary fiber to be, based on both physiological response and chemical identity.

It appears that we may be seeking a very elusive least-common denominator in such research.

The least-common denominator here is the fiber. There are different forms and different structures. They have different effects on human health that must be correlated with their different chemical compositions. It appears that we are still looking for that perfect analytical method for all forms of fiber.

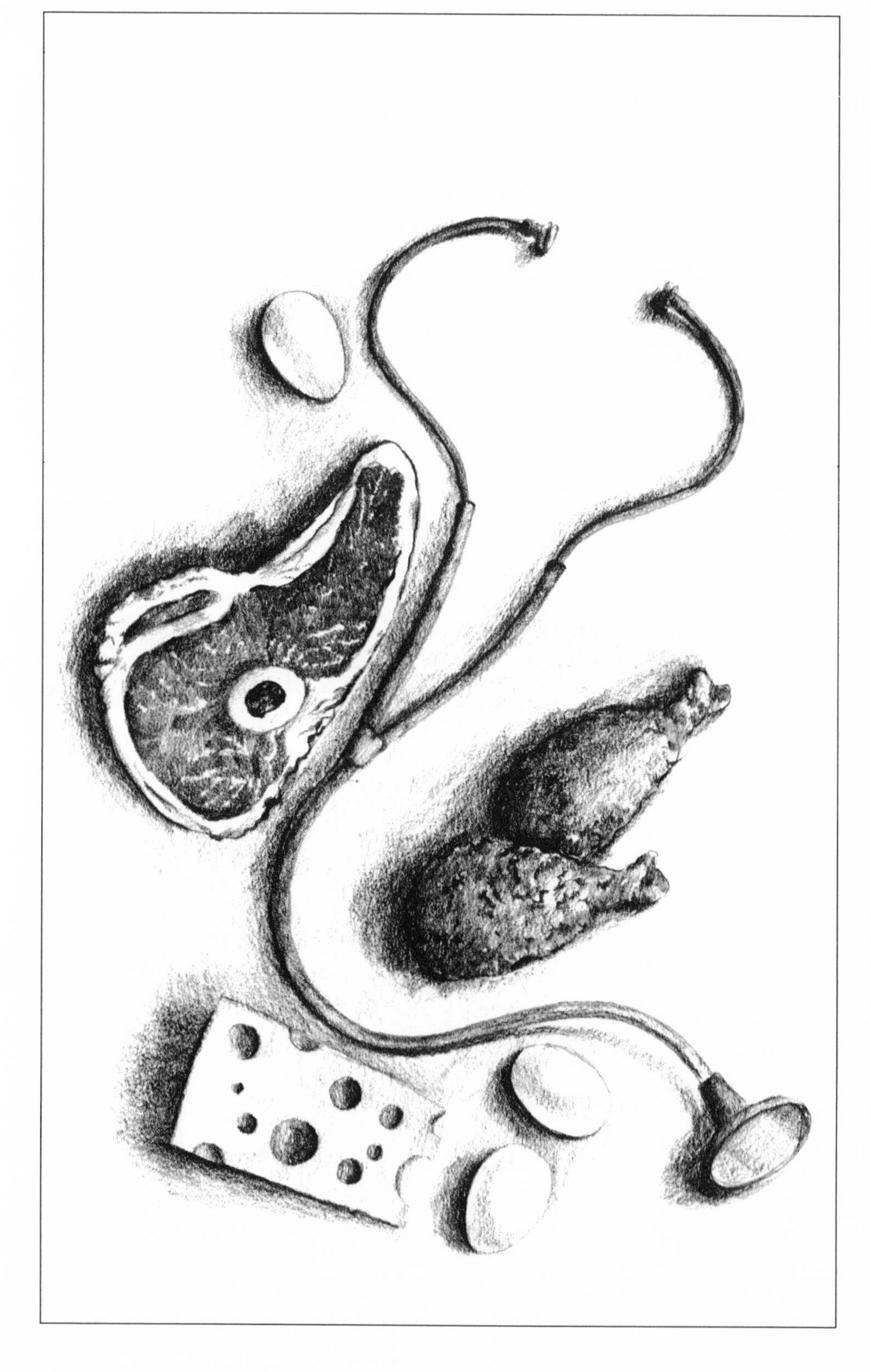

3

Fiber and Metabolism vs. Fat and Cholesterol

We hear so much about fats these days. Some fats, such as those containing mono- and polyunsaturated fatty acids, are considered good for humans. Others, with high levels of saturated fatty acids and/or cholesterol, are not. (*Saturated* means that the carbon atoms are surrounded by as many hydrogen atoms as is possible. *Unsaturated* means that there are fewer hydrogens around the carbons than there could be.)

The word fat means different things to different people, and most meanings are negative. To overweight people, fat means calories and virtually all foods containing lipids (fat, oil, glycerides, and cholesterol). To the average consumer, it is animal products like eggs, cheese, beef, pork, and lamb, which are also the high-quality proteins. To the physician, fat is the major cause of heart disease, vascular disease, and obesity. Lest we forget, however, some fatty acids are essential for good health. Moreover, most food fats and oils contribute to the pleasing flavor of cooked foods such as meats, sausages, pork products, fried foods, and roasted nuts. Without some fat these foods would lose their desirable flavor and not satisfy the appetite for long.

TRIGLYCERIDES AND CHOLESTEROL

Fats that are metabolized in the GI tract enter the bloodstream to be absorbed and subsequently utilized or deposited somewhere in the body. The fats in blood samples are referred to as serum triglycerides in clinical analyses. In general, the higher the dietary fat intake, the greater the potential for elevated serum triglyceride and cholesterol levels.

Glycerides are chemical complexes consisting of glycerol bound to one, two, or three fatty acids. Most of the fatty acids in edible oils and fats are 18 carbons long with lesser amounts of 16-carbon acids and 5–6% or less of all other fatty acids.

Cholesterol, by contrast, contains only 27 carbons but is a more complex chemical than the glycerides. The basic sterol structure has three six-carbon rings and a five-carbon ring joined side by side, sharing a two-carbon "side" at the junctions. The other carbons are present in a seven-carbon chain joined to a carbon of the five-membered ring and in three one-carbon methyl groups on different points in the six-carbon rings. As with the glycerides, some cholesterol is necessary for good health. It is present in all cells of the body. It is converted to hormones by the adrenal glands and the gonads, and into bile acids by the liver. The liver also synthesizes much of the cholesterol required by the body. This synthesis can be influenced by dietary fat or cholesterol, some forms of dietary fiber (as will be shown later), stress, and smoking.

The liver also converts cholesterol into the bile acids needed for proper emulsification of dietary fat so that the fat can be digested by the pancreatic enzyme, lipase, and subsequently metabolized and absorbed by the body. In response to a fatty meal, the gallbladder excretes stored bile, which contains the bile acids as salts, cholesterol, and phospholipids (another form of emulsifier) in a delicate ratio. Any major change in the balance among the three compounds can cause the formation of gallstones, which are made of crystallized bile salts and/or cholesterol. As bile acids are excreted into the GI tract, some are lost in the feces, but most are reabsorbed for return to the gallbladder. If the body's supply of bile acid is too low, the liver makes more from cholesterol circulating in the blood.

Cholesterol consists of both "good" and "bad" forms: HDL (high-density lipoprotein) and LDL (low-density lipoprotein). HDL is referred to as the good cholesterol because it confers some protection from heart disease. Increased levels of blood HDL have been associated with decreased risk of heart attacks, possibly by preventing the buildup of cholesterol deposits in the arteries. HDL carries about 20% of the body's cholesterol. The noted Framingham study of the mid-1970s showed that, in men, as the HDL level in blood fell below the average value, the risk of heart attack increased by as much as 25%. LDL, the bad cholesterol, carries the majority of total blood cholesterol, about 80%. Researchers think that LDL allows cholesterol to deposit in the arteries, and these deposits give rise to blocked arteries and lead to heart disease and strokes. Thus measurements of total blood cholesterol are more meaningful if broken down into the ratio of HDL to LDL. Because LDL contains most of the blood cholesterol, any significant drop in LDL will be reflected as a decrease in total cholesterol. Saturated fats in the diet tend to raise LDL cholesterol.

Until fish oil and omega-3 fatty acids became the latest rage, vegetable oils were our major sources of polyunsaturated fats. Fish oil supplements have also been shown to lower serum cholesterol and triglycerides in clinical studies. The well-reported comparison of coronary heart disease in Greenland Eskimos and Eskimos living in mainland Europe showed that those who ate fish oil fatty acids (polyunsaturated) had a lower incidence of heart disease. Nevertheless, the American Heart Association is still not convinced about the effectiveness of fish oil supplements; people who consume large amounts of fish generally eat less red meat, a major source of saturated fat, so some experts wonder whether the benefits are due solely to the use of fish oil or instead to a lower consumption of saturated fat.

Regardless of what it does for health, the fish oil bonanza is good for the economy. Fish oil costs about $300 a ton by the tank car. But fish oil capsules sell for 9–25 cents each, which translates into a retail value of $81,000–$224,000 per ton and $200–$500 a year for believers who take them regularly. Eating the fish is just as efficient (perhaps more so; the Eskimos ate fish, not capsules), and a great deal cheaper.

What about vegetable oils? In general, they are unsaturated, except for palm and coconut oils and fully hydrogenated oils used in shortenings. No vegetable oils contain cholesterol. Olive oil has the highest content of oleic acid (a monounsaturated fatty acid), followed by canola (the new Canadian rapeseed oil), peanut, and a newly developed high oleic acid sunflower seed oil. (Oils containing high amounts of oleic acid are reported to have beneficial effects on the heart.) The other edible polyunsaturated vegetable oils come from corn, soybeans, sesame seeds, safflower, and cottonseed. All of these oils are good sources of polyunsaturated fatty acids for edible uses.

EFFECTS OF DIETARY FIBER

Some forms of dietary fiber can affect cholesterol and bile acid levels in the body and the digestion of fat by lipase, with varying effects on human health.

Atherosclerosis and Cholesterol

Atherosclerosis is probably the major cause of heart disease in the Western world. It accounts for half of all deaths past the age of 45. The disease is manifested by the buildup of fatty deposits called plaques on or in the inner membrane of the arteries that carry blood throughout the body. Cholesterol is the major component of plaques, along with some fat and protein.

The level of cholesterol in the arterial walls increases with age; it is lower in young children. During life, all kinds of substances—sugars, amino acids, vitamins, minerals, and lipids—are continuously passing in and out of the arterial walls. Lipids include the fatty acids from digested triglycerides and cholesterol being transported by lipoproteins in the blood. If cholesterol is allowed to build up, it may become part of the plaques, and if this process continues unchanged, the inner diameter of the arteries becomes narrow and clogged; this condition leads to various forms of heart disease.

Dr. David Kritchevsky and his co-workers at the Wistar Institute in Philadelphia, PA, have conducted research on the relation of diet and atherosclerosis for more than 15 years. They found that heart disease correlated well with elevated levels of serum cholesterol, so that diet changes first concentrated on limiting dietary fat intake, especially saturated fat and cholesterol. However, some factor other than just fat in the diet seemed to be responsible for atherogenicity. Kritchevsky suggested in 1983 that the type of dietary fiber had an effect, because of its influence on the excretion of bile acids. Some types of fiber, such as pectins and gums, lower serum cholesterol, but wheat bran does not. High-fiber diets promoted the excretion of more bile acids, fat, and sterols. Bile acid excretion is the main pathway for elimination of internally produced cholesterol, and its excretion can lower serum cholesterol.

In the 1970s, Story and Kritchevsky reported lower serum cholesterol levels and degrees of atherosclerosis in animals when wheat straw, alfalfa, or peat was substituted for cellulose in the diet. This effect, they found, was related to the binding of bile acids. Wheat brans and purified fibers like cellulose did not bind bile acids. But lignin-containing fibers like alfalfa and wheat straw and hemicellulose-containing food residues from apple, celery, lettuce, potato, and string beans did. This finding was confirmed when, in 1982, Vahouny reported that wheat brans do not have significant effects on serum lipids or atherosclerosis. Dietary fibers from fruits, vegetables, and legumes showed much better response. The gel- and mucilage-type fibers, such as pectins and gums, decreased serum triglycerides and cholesterol and increased excretion of bile acids and cholesterol via the feces.

It has been known since the 1960s that oats may be the best of the cereal grains for lowering cholesterol levels, because of the presence of gums. In experiments with rats, DeGroot and co-workers replaced dietary starch with various fibers from cereal grains, then measured serum cholesterol levels. Rolled oats reduced cholesterol by almost 73%. Fibers from rice, barley, and whole wheat bread reduced cholesterol levels to some degree, but not as much as oats did. In 21 human volunteers, similar reductions were obtained; with rolled oats in the diet, cholesterol decreased, on average, from 251 down to 223 milligrams per 100 milliliters of blood after just 3 weeks.

Our own studies on the effects of purified rice hemicellulose fiber on fecal and blood lipids of rats confirmed DeGroot's results (Normand et al., 1984). Diets contained either 3% purified rice hemicellulose or cellulose (Alphacel) added to the control diet. Of the three diets, 3%

rice bran hemicellulose produced the highest total fecal lipids, but this result was due to increased excretion of triglycerides, not cholesterol. Neither the rice hemicellulose nor the cellulose increased fecal cholesterol excretion compared to the control diet. Analysis showed slight decreases in total blood lipids, but not serum cholesterol, of fiber-fed rats. Though rice hemicelluloses do not lower serum cholesterol levels, they do appear to have favorable effects on serum triglycerides by virtue of their binding to bile acids, as we will see later.

So there is now ample evidence that *some forms* of dietary fiber can reduce serum cholesterol levels and the subsequent development of atherosclerosis. But other factors also can affect cholesterol. For example, there is considerable evidence that dietary protein plays an important role. Researchers have shown strong correlations between intake of animal protein and mortality from heart disease, serum cholesterol, and atherosclerosis. Blood cholesterol levels in humans have been significantly reduced by substituting soybean protein for animal protein in the diet. Serum cholesterol and atherosclerotic lesions were increased in rabbits fed animal protein and reduced with soybean protein. And cholesterol levels can be lowered by several other factors too: decreasing fat intake (especially saturated fat and cholesterol), proper exercise, limiting alcohol consumption, and avoiding smoking.

Triglycerides and Binding of Bile Acids

We have noted that some forms of fiber lower cholesterol levels in the blood by binding bile acids and excreting them in the feces. However, some forms of fiber can bind to bile acids and not have any effect on cholesterol levels. Bile is released from the gallbladder into the intestinal tract in response to a meal. After the bile acids have done their job of emulsifying fat for digestion, the majority are reabsorbed by the body and returned to the liver. Those lost in the feces can be replaced by synthesis from serum cholesterol circulating back to the liver.

What about the binding of bile acids or salts by fiber? Can this be beneficial even if it does not lower cholesterol levels? It should be remembered that the major role of bile acids is not to lower blood cholesterol but to emulsify food fats for the body to break down for digestion and absorption. The other effects are important but secondary.

Kritchevsky and his co-workers have done a great amount of research on fiber–bile acid interactions. He varied protein, carbohydrate, and fat contents in the diets of rats and fed them alfalfa or cellulose as fiber. Alfalfa increased the excretion of cholesterol much more than did cellulose. In a relative comparison of binding of nine bile acids by four different fibers (alfalfa, wheat bran, cellulose, and lignin), they found different rates of binding between the different bile acids and between all fibers. Cellulose and cellophane bound little of each of the bile acids. Bran bound high amounts of four bile acids and medium to low amounts of the others. Alfalfa and lignin bound the most and thus were the most effective fibers to bind bile acids.

When the fiber from barley or oat husks, corn meal, apples, Brussels sprouts, carrots, pears, and turnips was tested, the same kinds of differences in binding affinity for bile acids was noted, confirming that fibers from different plants have different binding properties. In 1978, Kern and his co-workers studied the binding of bile salts to fibers from apples, corn, kidney beans, lettuce, potatoes, and string beans and found the same results. Highest binding was by string beans, followed

by celery, corn, potatoes and lettuce (which almost tied), kidney beans, and apples. They concluded that although very large amounts of some foods are required to bind a high amount of bile acids, relatively small amounts of kidney beans, potatoes, string beans, and corn can bind 1 gram of chenodeoxycholic acid, one of the major bile acids. This observation may have some clinical significance. (Isn't this what we said before? Eat a *balanced* meal consisting of more than one vegetable, some fruit, limited saturated fats, and a cereal grain product, preferably whole grain, and the results should be beneficial.)

In our studies of binding of bile acids by rice hemicelluloses (fractions), we found striking differences between the different varieties of rice (Normand et al., 1979, 1981, 1987). Long-grain rice (preferred for Southern cooking because it cooks as separate grains) was not as effective as medium-grain rice (the favorite for Oriental cooking because the grains cling together and are better for eating with chopsticks). In Table 1 we see that rice bran hemicelluloses (fractions) bind different

Table 1. Binding of Four Bile Acids by Hemicelluloses from Rice and Wheat

Source	*Cholic*	*Glycocholic*	*Taurocholic*	*Glycotaurocholic*
Bran fraction				
Long-grain rice	17.4	11.1	17.7	11.6
Wheat	17.8	10.7	17.3	21.1
Medium-grain rice	25.7	40.3	21.9	28.4
Whole grains				
Long-grain rice	44.2	25.6	24.5	35.9
Wheat	14.8	14.1	4.2	16.9

Note: All values are given as percent bound.
Source: Data are from Normand et al., 1979.

amounts of cholic, glycocholic, taurocholic, and glycotaurocholic acids, but hemicellulose from medium-grain rice bound greater amounts than both long-grain rice and wheat (whole grains), under conditions simulating those in the GI tract. These studies also showed that higher levels of hemicelluloses added to constant amounts of the bile acids bound 21–34% more than did lower levels. That result indicates that the amounts of bile acids bound are dependent upon the amount of hemicellulose, or fiber, present; the more fiber present, the more bile acids that are eventually bound. This same effect was noted by Story and Kritchevsky in 1976 and 1978.

♦

This brief review of the voluminous research literature on the effects of dietary fiber on fat and cholesterol metabolism and atherosclerosis shows that there are beneficial effects. However, it is evident that no single statement can describe these benefits. Each fiber has its own properties. Some lower blood cholesterol levels. Some lower blood triglycerides but not cholesterol. Some lower both. Some bind bile acids, and some don't. The type and amount of fiber must be identified before any predictions can be made. We now recognize, however, that those fibers that alter lipid metabolism can be beneficial to health and lessen the risk of heart disease.

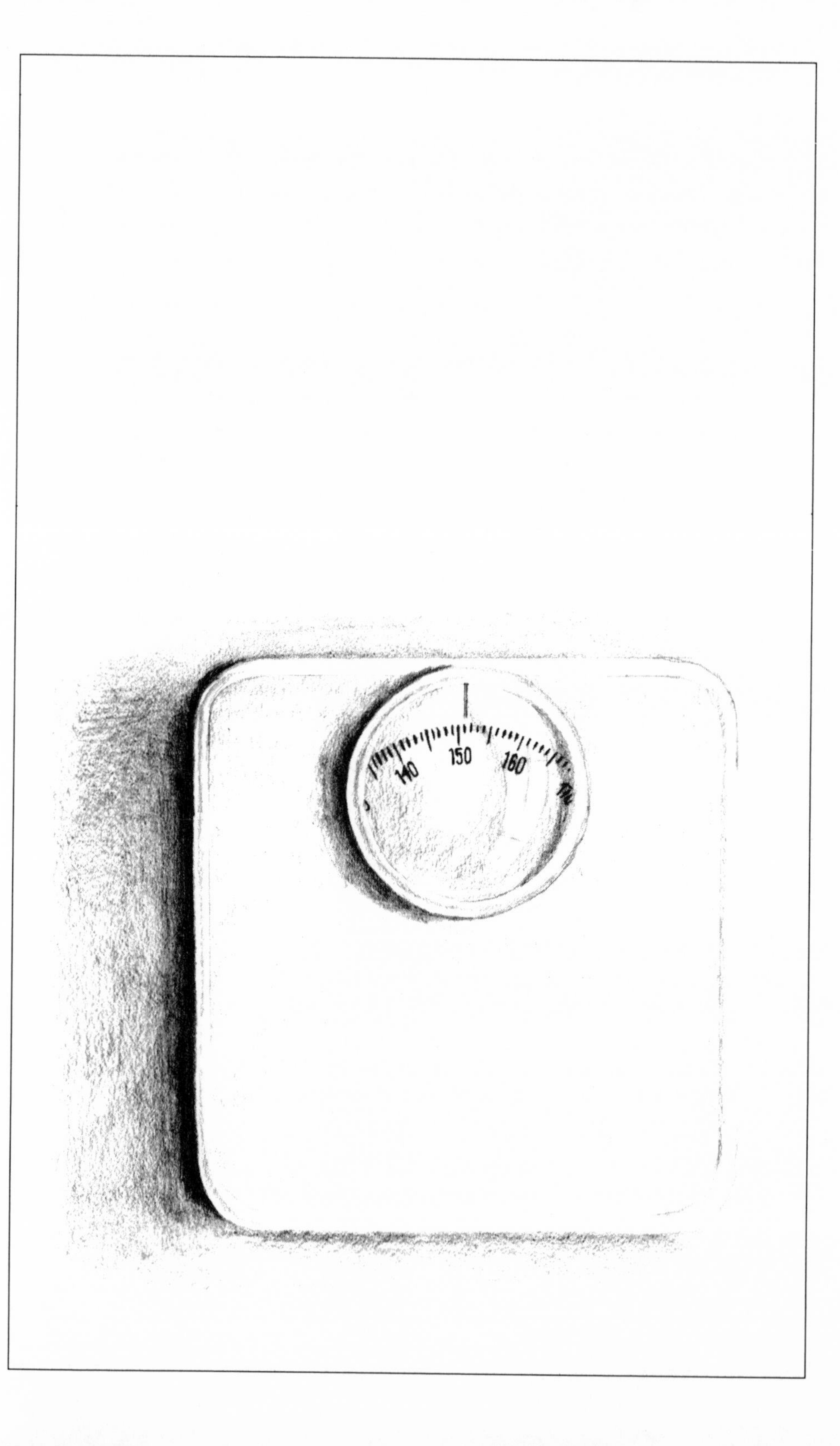
140
150
160

4

Can Fiber Fight Fat?

To control weight, two factors are required: determination of the overweight person and a nutritionally adequate diet that provides the optimum amount of calories needed for the desired goals. Dietary fiber can play an important role in controlling obesity. Dietary fiber helps people feel full and have fewer hunger pangs, partly because the absorption of water by the fiber promotes the satisfied feeling. Calorie contents of fiber-containing foods are generally lower than equivalent weights of protein or fat-containing foods. Cereal bran products, fresh fruits, and vegetables are usually low in fat and protein but high in complex carbohydrates. Fruits and vegetables are also usually high in water and low in solids, so they allay feelings of hunger while providing few calories. The higher the fiber content, the lower the caloric intake, which is what is required to lose or control excess weight.

Owen and Cotton in 1982 described a study conducted with 16 overweight male volunteers at Michigan State University. Each day for 8 weeks the men ate four slices of bread with each meal. Half ate regular bread and half ate a high-fiber bread containing wood pulp (cellulose). Both groups lost weight, but those eating the high-fiber bread lost an average of 19.3 pounds, compared to 13.8 pounds for those eating regular bread. Those eating the high-fiber bread also had less between-meal hunger; between-meal hunger can be a big problem for dieters trying to lose weight.

Although termites have eaten wood pulp for centuries, humans have only been eating cellulose processed from wood pulp or cotton since about 1950. Microcrystalline cellulose (Alphacel) is added in minute quantities to selected food products, such as ice cream, salad dressings, toppings, and puddings, as a stabilizer. It prevents ingredients from separating and ice crystals from forming. In larger quantities cellulose is being added to "diet" breads as a fiber source and for calorie control. Bakers are able to claim that the bread has 50–400 times more fiber than regular bread, and 25–30% fewer calories. Although this cellulose is different from the fibers in bran cereals, fruits, and vegetables (*see* Chapter 2), it does hold a great amount of water and it has no nutritive value for humans. It can, therefore, be an agent for obesity control and colonic stimulation if it is not consumed in excessive amounts that could affect mineral balances.

EFFECT OF FIBER ON FAT EXCRETION

To determine if the binding of bile acids affected lipid metabolism in living organisms, we examined the feces and blood of rats fed rice hemicelluloses or cellulose (Alphacel) (Normand et al., 1984, 1987). As already noted, neither fiber increased cholesterol excretion or decreased blood cholesterol levels, but both increased excretion of total lipids, including undigested triglycerides. Table 2 shows that rats fed rice hemicelluloses excreted the highest amount of total lipids, most of this as undigested triglycerides. Looking at lipids in the blood, rats fed the rice hemicellulose diet had the lowest total compared to the control and the

Table 2. Effects of Rice Hemicellulose and Cellulose on Total Lipids in Feces and Blood of Rats

	Type of Diet		
Fraction Analyzed	*Control*	*3% Rice Hemicellulose*	*3% Cellulose*
Feces total lipids (five rats pooled)	472.1	610.1	464.0
Blood total lipids (per 100 grams of dried blood)	608.4	549.8	575.2

Note: All values are given in milligrams.
Source: Data are from Normand et al., 1984.

cellulose diets. This result makes sense because fat (triglycerides) that is excreted is by definition not absorbed into the bloodstream and retained by the body. After absorption from the intestines, the free fatty acids and monoglycerides are converted back into triglycerides or other fatty structures to be transported by the blood to other parts of the body. Excessive amounts of triglycerides are dropped off and stored in the body's fat depots, or adipose tissue, where they increase body weight and waistline (plus other noticeable places).

How can fiber prevent these excessive deposits of fat in the adipose tissue of the body? The pancreatic enzyme (lipase) that digests the triglycerides must have the water-insoluble fat in emulsified form for proper digestion. There have been several theories for the action of dietary fiber on pancreatic lipase activity. Some propose that fiber binds to the enzyme. Others propose that unpurified fibers, such as those found in balanced meals, contain inhibitors that alter pancreatic secretion or absorption of digestion products.

We examined the effects of rice bran hemicellulose on pancreatic lipase activity under laboratory conditions that simulated a living organism (Normand and Ory, 1984). These results (Table 3) show that fiber has an indirect effect on the enzyme. As we expected, without added

Table 3. Effects of Rice Bran Hemicellulose and Two Bile Salts on Pancreatic Lipase Activity

Reaction Conditions	*No Bile Salts*	*Cholate*	*Taurocholate*
Control	0	—	—
Bile salts alone	—	12.0	9.0
Hemicellulose–substrate premixed before fat added	—	10.0	16.5
Hemicellulose–bile salts premixed before fat added	—	4.5	5.0
Hemicellulose alone	4.0	—	—

Note: All values given are lipase activity in microliters per minute.
— indicates none determined.
Source: Data are from Normand and Ory, 1984.

bile salts there is no lipase activity on the triglycerides. With bile salts, such as sodium cholate or sodium taurocholate, the enzyme's activity increases significantly. If the hemicellulose fiber was mixed with fat before adding bile salts, lipase activity was not reduced significantly. But if the hemicellulose was mixed first with the bile salts before adding the fat and the lipase, enzyme activity was greatly reduced. These results indicate that bile salts are essential for emulsification of the fat prior to the enzyme breaking it down. If the bile salts are not available, less fat is emulsified, lipase activity is generally reduced, and the fat cannot be properly digested. Decreased lipase activity in humans may result in less absorption and possibly increased excretion of fat in the feces. These effects appear to have occurred in the rat feeding tests described in Table 2.

For the overweight individual, such events would mean that meals containing high-fiber foods, e.g., whole brown rice or some form of rice bran, would prevent the fat in such meals from being completely absorbed. Because the fat would not be properly emulsified, less lipase activity would take place, and the undigested fat would be excreted in the feces. If caloric intake is limited or decreased, and some form of exercise is followed to "use up" calories, the net result could be a gradual weight loss for the person.

DIETING

There have been more diets and gimmicks to lose weight than can be discussed here. The only good diet is one that produces good results and maintains weight loss. Because everyone's body is different, a good diet for one person may not work for another. A good diet should be easy to follow and not be monotonous. It should consist of simple foods that are readily available and have sufficient good taste to be palatable. The regimen should include some type(s) of exercise to use up calories, improve muscle tone, and strengthen cardiovascular tissues (heart and blood vessels). Diet pills, diet drinks, and diet gimmicks cannot take the place of a balanced but limited-calorie meal. If hunger pains get severe between meals, snack on low-calorie items like fresh

fruit, carrot or celery sticks, other vegetables, popcorn without butter or salt, or a handful of cereal instead of candies, chips, nuts, or bakery confections.

(Even in South Louisiana where Cajun and Creole cooking has been a tradition, people are beginning to diet. I heard two Cajuns discussing their new diets. Said one,

"I never eat foods high in fat, sugar, and chemical additives now. I don't eat meat, eggs, dairy products, chicken or fish anymore."

"Man!" said the other, "How do you feel?"

"Pretty doggoned hungry, my friend," was his reply.)

A diet that cuts out virtually all foods is not a diet, it is a crash course in malnutrition.

Every time you eat 3500 calories more than your body needs for maintenance, you add one pound of body fat. Conversely, when you reduce caloric intake and/or increase exercise to use up 3500 calories, you lose a pound.

In place of high-fat foods, substitute more complex carbohydrates (whole-grain products, dried beans and peas, and fresh fruits and vegetables). That is what distance runners eat before a big race. Some of the recipes in Chapter 11 can serve as suggestions for low-fat, high-fiber foods. There are so many ready-to-eat breakfast cereals available today that breakfast can be the most important source of fiber each day. Data

from the 1986 National Food Consumption Survey showed that people who eat cereals for breakfast are less likely to skip breakfast than those who eat other breakfasts that require some sort of preparation.

One can lose weight or maintain body weight by combining exercise with a balanced, limited-calorie diet. (Contrary to what so many young people think today, a balanced meal is not a hamburger in each hand.) Despite what we know today, however, there is still no "ideal diet" for each person because of differences in age, physical condition, general health, body size, amount of physical activity expended, illness, and pregnancy. The tables for proper weight are generally averages or ranges.

Some general dietary guidelines to help maintain good health and well-being through proper eating habits are as follows:

1. Eat a variety of foods containing sufficient complex carbohydrates and fiber.
2. Avoid eating too much fat (especially saturated fats and cholesterol), sugar, and salt (sodium).
3. Maintain a desirable weight by including some exercise or physical activity with the improved eating habits.
4. If you drink alcoholic beverages, do so only in moderation. It's probably best to avoid them altogether.

EXERCISE

Crash diets will not keep excess weight off; only a balanced, reduced-calorie diet accompanied by a regular, proper exercise program can take off pounds and keep them off after a desired weight is reached. Regular exercise will also condition the body, the heart, and the muscles; keep the body alert; reduce stress; and regulate the appetite to make the person less likely to eat excess calories the body doesn't need.

Use up extra calories by following a good exercise program. It doesn't have to include running, jogging, or expensive exercise equipment. Walking is excellent exercise; you can use up to 2000 calories (the amount estimated for maximum cardiovascular benefits) by walking 20 miles per week at a brisk pace. Aerobic exercises that increase the heart and breathing rates are good for strengthening the heart muscles and blood vessels, lowering blood pressure and blood sugar, and strengthening bone and muscle. These exercises can be done at home. Remember, however, to have a complete physical examination by your physician first. Once your doctor has confirmed that it is safe for you to do so, you can begin a program of exercise *within a safe range of your heart rate* (forget the "No Pain No Gain" theory). As you improve you can move up gradually to a set program of 20–30 minutes each day, 4–6 days per week.

Medical experts agree that exercise can help control weight, handle stress, aid sleep, increase energy levels, and improve blood circulation throughout the body. In turn, the heart, lungs, and muscles function more efficiently all day. According to the American Heart Association (AHA), the average heart rate is 70–75 beats per minute compared to 40–45 beats per minute for a well-conditioned person. This means that the heart of a sedentary (nonexercising) person works harder and may pump up to 36,000 times more per day than the heart of a person who exercises and is in good shape. The AHA recommends a lifetime commitment to some type of exercise program as a means of reducing the risk of coronary heart disease—along with eating a prudent diet low in fat and high in fiber. The AHA general program for monitoring heart rates during exercise is listed on page 42.

AHA Exercise for a Healthy Heart

Age (years)	*Average Maximum Heart Rate* (100%)*	*Target Zone* (60–75%)*
20	200	120–150
25	195	117–146
30	190	114–142
35	185	111–138
40	180	108–135
45	175	105–131
50	170	102–127
55	165	99–123
60	160	96–120
65	155	93–116
70	150	90–113

**Beats per minute*
Source: The Clarion Herald, New Orleans, LA, 6 Aug. 1987, sec. 3.

Thus, for a 20-year-old person, the maximum heart rate would be 200 and the target rate 120–150 beats per minute, i.e., 60–75% of the maximum of 200. Age is not a barrier if your physician finds no reason you should avoid such exercise. You should think young; aging is for wines. After getting used to an exercise and diet control routine, you will begin to feel better and have less anxiety and depression. Of course, you can become tired and hungry after a period of exercise. The key here is *not* to relieve this hunger with a milk shake, a beer, or a big sandwich. If you must eat or drink something, try fruit juice, a diet drink, or a piece of fruit.

Despite the growing interest in running and other kinds of physical exercise, a large percentage of the population still does not engage in some form of physical activity, at least not enough to stimulate the cardiovascular system and burn off excess calories. The lack of exercise and the consumption of highly refined foods lead to obesity or various forms of heart disease. Those who do not exercise should benefit from an increased intake of dietary fiber, especially important for the group over age 50. It takes a lot more effort to start exercising tired bones. But you can set aside a portion of each day, preferably in the morning before breakfast, to do about 20–30 minutes of some form of aerobic exercise (walking or jogging outdoors, calisthenics or vigorous in-place exercises indoors) to set up your heart rate. If the weather is bad, or you do not feel like going outdoors, there are many forms of exercise that can be done indoors—depending upon your age, health, physician's recommendation, and physical condition. For example, if you can get down onto the floor (and get back up again afterwards), lie on your back and do "deep knee bends" by bending your knees, raising your legs off the floor, and pulling the knees toward the chest. Lower your feet slowly to the floor and repeat the exercise a set number of times. Another simple but effective exercise is to stand straight, grasp the back of a chair, and alternate lifting each knee to the waist or beltline. If you are confined to bed or to a chair, try outstretched-arm exercises: lift them up, down, side-to-side. Bend the torso and move your legs and feet. To increase the challenge, try this with a book or a can of beans in each hand. Start with a few of these exercises and increase the number as you become accustomed to the increase in activity. The activity table that follows suggest other forms of exercise that can help use up calories and tone up old, tired muscles. Once your body has adjusted to the new routine and your stiff muscles have relaxed, exercise becomes an easy task that will actually make you feel better and more alert for the day's activities.

Last but not least, don't sit on a couch in front of the television set right after eating. Take a 15–20-minute walk, or do some moderate form of exercise to help you to relax and settle the meal. (Never engage in strenuous physical exercise soon after eating, especially if you're obese. The extra fat makes the body work harder, and it insulates the body and thus makes it harder to diffuse the heat generated. If the fat is around the heart, it can even kill you. The safe approach is to undergo a complete physical examination by a physician before beginning the exercises.)

The New Orleans Dietetic Association listed the approximate calories used in selected activities in *The Times—Picayune*, page E-10, May 5, 1988.

Activity	*Calories per hour*
Strolling or standing	120–140
Walking (2 mph[a])	150–240
Walking (3.5 mph)	300–360
Walking (5 mph)	420–480
Housework	
Mopping, vacuuming	240–300
Scrubbing floors	300–360
Raking leaves, hoeing in garden	300–360
Digging in garden	360–420
Bowling, golfing and pulling a golf cart	240–300
Table tennis, golfing and carrying the clubs	300–360
Cycling (10 mph), roller skating, dancing	360–420
Cycling (12 mph)	480–600
Tennis	480–600
Jogging (5 mph)	480–600
Running (5.5 mph)	600–660

[a]*mph is miles per hour.*

WHAT NOT TO DO

There is no such thing as a food that "burns off fat". Fat can only be burned off by using more calories than the body takes in via the food we eat. Fad diets waste money, not pounds. Beware of "super" pills, grapefruit, banana or rice diets, skipping meals, and starch blockers (which may contain substances that inhibit other nutrients too). Dieting succeeds best when weight loss is gradual, not by crash diets.

Skipping breakfast as a means of losing or controlling weight is not advisable. Contrary to certain beliefs, skipping breakfast does not result in increased weight loss. It only increases hunger and lowers energy, work performance, and efficiency. Breakfast may be the most important meal of the day because the body has not had food since the evening meal of the previous day, 12–15 hours earlier. Blood sugar level is low, and the body needs a new source of energy to get started for the day's duties. If the energy level is not raised through a good breakfast, the individual cannot function at the best level, children can't concentrate and learn as well, and the individual's ability is impaired.

Fast food services are almost a way of life for many, but a meal containing a hamburger with lettuce, tomatoes, etc., French fried potatoes, and a milkshake will contribute almost 50% of the calories as fat, much of it as saturated fat. This amount is 20–30% higher than the amount of fat calories recommended by the American Heart Association. More will be said of fat as one of the major nutrients in Chapter 8.

Another thing for the obese or overweight to remember is that *ALL* fats, both saturated and unsaturated, contribute 9 calories per gram of fat (compared to 4 calories per gram of protein and carbohydrate) when consumed, metabolized, and absorbed. A person trying to limit calories should not feel that avoiding saturated fats is enough and that consuming large amounts of unsaturated fats and vegetable or fish oils is all right.

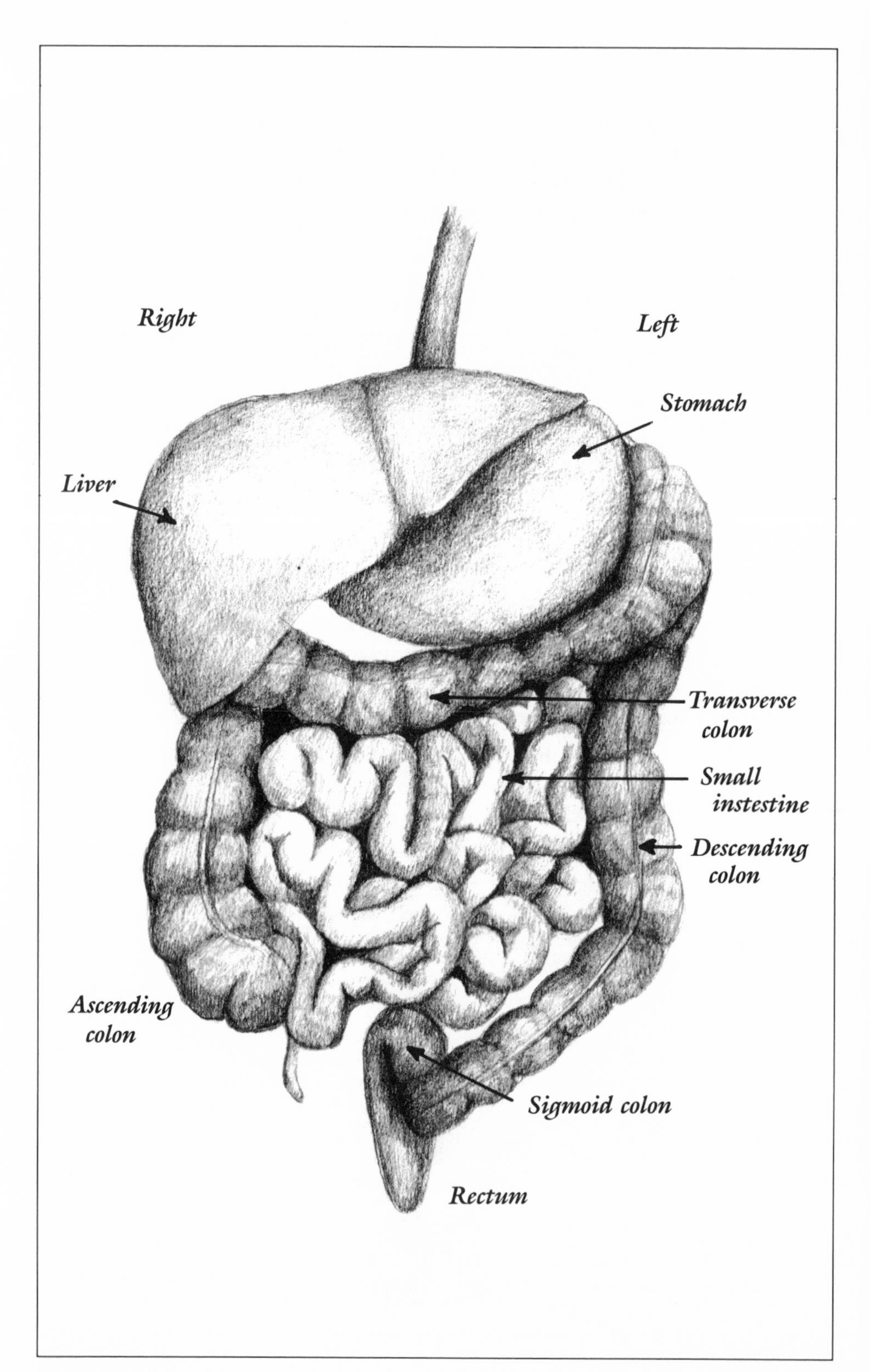
Right
Left
Stomach
Liver
Transverse colon
Small instestine
Descending colon
Ascending colon
Sigmoid colon
Rectum

5

Fiber and the Big "C"—Constipation

Of all the benefits attributed to dietary fiber, improvement of bowel function and its subsequent effects on various colonic functions is probably the most widely publicized. Dr. Denis Burkitt, who was studying colonic diseases in rural African and Western populations barely two decades ago, noted that the diets of these people correlated well with their incidence of colon diseases and bowel habits ("regularity", as it is known in television commercials). Dr. Burkitt noted that rural Africans ate coarse grains, beans, and vegetables with little or no animal protein. Their bowel movements were easy, with soft, bulky stools, and they did not suffer from constipation, hemorrhoids, diverticular disease, colon cancer, and other disorders that were common in European populations that consumed high animal protein diets with milled wheat products and fewer vegetables. He concluded that the increase in constipation and colonic diseases of the Europeans correlated with the decline in consumption of coarse grains and increased consumption of milled wheat products near the end of the 19th century.

Fiber, especially hemicelluloses, pectins, gums, and hydrocolloids, seems to hold water in the large intestine, and, as Dr. Burkitt observed, the greater the amount of water, the softer the stools that are excreted. High fiber also promotes faster transit time through the intestines. Africans consume about 25 grams of crude fiber per day; the average American consumes 5–10 grams per day. Dr. Burkitt theorized that a low-fiber diet slows down the transit time, which means extra pressure is required to move the feces through the intestines. This need results in thicker intestinal walls, eventually causing extreme pressure in localized areas, popping through the muscle membranes, and the formation of diverticular “pockets”: small areas of ballooning or pouching of the colon wall. The sigmoid colon empties in response to the amount of waste matter in it. Dr. Burkitt estimated that at least 200 grams (almost half a pound) of material must be present in the sigmoid colon–rectum area to cause the reflex action that empties it, i.e., the bowel movement. With less material, there is not enough bulk, and the person must strain to empty the bowel. The constant straining can lead to other disorders such as hiatal hernia, hemorrhoids, and even pinching off of small blood vessels going to the legs to cause varicose veins. Dr. Burkitt suggested that fiber, by speeding up transit time and increasing the water content of the feces, also had a positive influence on the microorganisms of the large intestine, which break down some forms of fiber and thereby increase the fecal bulk.

About $250 million per year is spent on laxatives in the United States and Britain because people do not consume enough fiber.

DIGESTION

Digestion is an extremely complex process, but it can be summarized briefly by reviewing the illustration at the beginning of this chapter. Food enters the stomach after chewing and mixing with saliva (which contains the enzyme amylase to start the digestion of starch). In the stomach, food is mixed with a small amount of hydrochloric acid to provide a proper pH for the protein-splitting enzyme, pepsin, and a gastric enzyme (lipase) that digests fat. As food passes out of the stomach into the upper portion of the small intestine, more enzymes secreted by the pancreas (trypsin and chymotrypsin to digest protein, amylase to digest starch, and lipase to digest fat) are added to the intestinal contents, along with bile salts from the gallbladder that emulsify fat for digestion. As these foods pass through the 20–21 feet of small intestine, the protein, starch, and fat are broken down to their basic units: the proteins into amino acids, the starches into sugars, and the fats into fatty acids. But dietary fibers pass through essentially unchanged. The amino acids, sugars, and fatty acids are absorbed into the bloodstream and go back into circulation to form new body protein, glycogen (a high-energy carbohydrate), and body fat. If lots of fiber is present, bile salts may be bound so that fat digestion is partially inhibited, as we saw in Chapter 4.

Undigested materials pass from the small intestine into the large intestine at the bottom of the ascending colon. For practical purposes the large intestine can be arbitrarily divided into the right side and the left side. The right side is essentially a fermentation tank (like a cow's rumen), and the left side is essentially a storage tank where water is removed from the feces and reabsorbed into the body. The right side, the ascending colon, is where intestinal microorganisms break down some forms of fiber. This breakdown increases the bulk of the stool. The bacteria multiply and produce some volatile fatty acids and gases: hydrogen, carbon dioxide, and methane. Normal peristalsis moves these materials through the colon to the left side, the descending colon, where some water is removed and storage takes place. If things are moving quickly (because of fiber's presence, for example) the stool loses less water, and fecal bulk and softness are increased.

Filling of the sigmoid colon–rectum area causes a reflex action that prompts the urge to defecate. With high water and stool bulk, the action is quick and easy. Where fiber content is low, transit time is decreased, and water retention is low, leaving hard, dry stools of higher density. Emptying the rectum then requires straining and pressure. Result: *Constipation!* If stools become compacted, harsh laxatives or an enema may be required to unpack the blockage.

CONSTIPATION

Constipation is generally considered a problem that comes with age, although many people under the age of 50 also suffer from this condition. Because of significant advances in health care, the size of the elderly population in the United States and the entire Western world is growing rapidly. Nutritional problems of the aged are also growing, and we are finding out that we have much to learn about geriatric nutrition. The geriatric age group is generally considered to be age 65 and older. After age 50–60 a gradual decline in physiological functions begins. This decline creates nutritional problems due to changes in health status, food intake patterns, reduced salivary secretions, dentition, and ease of chewing. To avoid discomfort or eating problems, the elderly may select foods because they are nonfibrous, soft, or sweet rather than for their nutrient value. They also consume less fluids. With decreased phys-

ical activity, gastrointestinal motility is altered, and constipation becomes more of a daily problem that many control with chemical laxatives rather than by increasing fiber and complex carbohydrates in the diet.

LAXATIVES

As Dr. Burkitt and others have shown, a diet containing adequate fiber can prevent most constipation. Our grandparents also believed that a diet high in roughage would promote "regularity" and prevent constipation. If the bowels became sluggish from constipation, laxatives were given "to clean out the system". The chemical laxatives (especially those containing phenolphthalein) generally did so—in fact, they felt like the intestinal walls were being scrubbed clean! If the rectum was packed hard, an enema containing warm soapy water was used to dislodge the packed wastes. Today the multimillion dollar laxative industry also includes nonsoapy enema "kits". They contain phosphate-buffered water in a soft plastic bottle that can be discarded after use. It seems like a lot of money is being flushed down the toilets of America.

Contrary to the advertising blitz by the pharmaceutical industry in magazines and on television, "gentle overnight relief" and "promotion of regularity" are nice phrases that can cause more constipation than they relieve if the products are continued too long. Many doctors warn that frequent use of laxatives can lead to dependence upon the drugs, and, ultimately, to serious disturbances in the GI tract. Dr. John Deaton, an Austin, Texas, physician said,

> They suppress the normal function of the intestines and they can cause real harm to your body. Diverticulitis, the painful inflammation of ballooned-out projections in the intestinal walls of many middle-aged people, as well as appendicitis and an irritable colon are frequently attributed to the use of laxatives. It may sound paradoxical, but the first step in getting rid of constipation is to throw out all of the laxatives in the house. You'll be surprised how fast your gas pains, irritability, and indigestion will disappear after you stop taking them.
>
> (Dr. Deaton was quoted in a 1976 magazine article by J. Ellis.)

A chemical laxative that "cleans out the system" removes not only the stored waste matter in the rectum, but also cleans out most of the soft, semi-liquid waste in the colon well above this point. The result: there is no reflex urge, the nerve impulse to the brain that says, "I'm full; empty me," for the next day or so. People mistakenly feel that constipation still exists and take more laxatives, which short-circuit the urge, and the cycle of laxative dependency starts.

DIETARY FIBER

Dietary fiber is Mother Nature's own laxative. It comes in many forms and colors, many sizes and shapes, and the containers are not glass or plastic. It is contained in the foods we eat: cereals, fruits, and vegetables. Although daily habits, stress, lack of physical activity, and illness can affect bowel habits, proper diet can play a very important role in normalizing bowel functions.

Diverticular disease is one of the most common results of continued constipation, and it is most common in the elderly of Western societies. It is virtually unknown in the less-developed countries, where little or no processed foods are eaten. Older bodies require fewer calories to function properly so that, to avoid the old age weight gain so common in Western populations, it is essential that caloric intake—

especially fat—be reduced accordingly. By replacing fat with protein or complex carbohydrates, caloric intake can be reduced. If the complex carbohydrates contain a good portion of dietary fiber, the caloric intake can be reduced significantly and still provide the roughage that can prevent constipation and reduce diverticular disease and the associated disorders that result from this condition.

The parts of the body that are not composed of fat consist of about 70% water, so water is a major nutrient. It acts as a lubricant in saliva. It functions in digestion and absorption of nutrients in the stomach and upper intestines. It dissolves other nutrients and waste products of digestion for circulation by the blood and for excretion in the urine and feces. When water is absorbed in the large intestine, the result is softer stools, less straining and pressure within the colon, less constipation, and less diverticulosis. Because of this property, a great deal of research has been devoted to the water-holding properties of various foods. Several methods have been used but, in general, they all involve taking a weighed amount of the fiber or fiber-containing food, placing it in contact with an excess amount of water for some period of time, then measuring the water uptake by one or more methods. Another method used is to feed small animals (usually laboratory rats or mice) a specific fiber, collect the fecal pellets frequently to get fresh weights, then dry the pellets carefully and weigh them again. The difference in the weight of fresh and dry feces is that of the water that was taken up or held in the feces.

In our rat-feeding studies mentioned earlier (Normand et al., 1984), we compared the water-holding capacity of rice bran hemicelluloses and cellulose (Alphacel) with the weights of fecal pellets. Feces were collected each day for 7 days and weighed. Each rat eating a fiber diet showed significant increases in wet fecal weight compared to those eating the control diet. When fresh weights for each group were pooled, the values were 8.87 grams for the controls, 13.97 grams for the rats on rice hemicellulose, and 21.14 grams for the cellulose-fed rats. This result illustrates that both fibers increase the water content of the feces, but the cellulose, which is not broken down by colon microorganisms, increases fresh fecal weight more than the hemicelluloses, which are partially broken down by the colon bacteria. However, when the pellets were dried, the true water contents were obtained. Table 4 shows that

Table 4. Effects of Rice Hemicellulose and Cellulose on Rat Fecal Output and Fecal Water Content

	Type of Diet		
Substance Weighed	*Control (No Fiber)*	*3% Rice Bran Hemicellulose*	*3% Cellulose*
Feces, wet weight (grams)	1.3	2.0	3.0
Feces, dry weight (grams)	1.1	1.7	2.7
Water content (%)	11.8	15.5	10.3
Feed consumed (grams)	52.3	54.6	56.4

Note: Five rats per day were studied.
Source: Normand et al., 1984.

both fiber-fed groups had higher wet and dry fecal weights than the no-fiber control, but the water content of the hemicellulose-fed rats was the highest. When the average consumption of the three groups was calculated, the cellulose-fed rats ate more than the others, a finding that may help explain the higher fecal weight collected.

Other researchers have also found that increasing the fiber contents of the diet caused rats to compensate for the lower energy concentration by consuming more feed. In other words, both kinds of fiber, cellulose and hemicellulose, may increase fecal output by increasing both water content and solid residue that contains some undigested dietary fiber.

Dr. Martin Eastwood investigated the ability of various plant sources to increase stool weight. He reported in 1977 that carrot was most effective, followed by cabbage, sugar beet pulp, peas, and wheat bran. He also reported that 16 grams of bran daily could almost double the stool weight after 3 weeks on the diet. Cellulose was as effective as

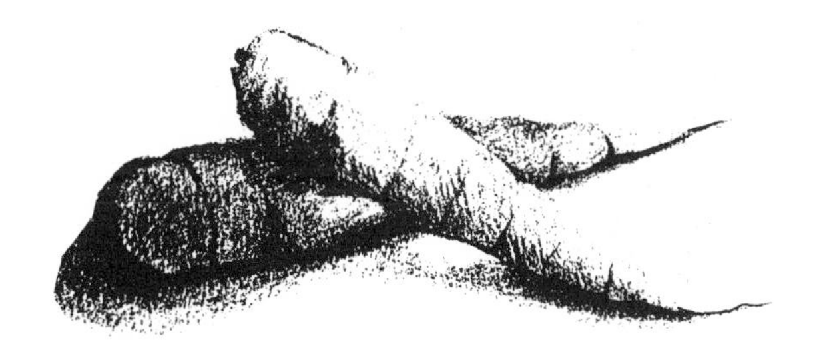

wheat bran. He did note, however, that the effects of cereal bran on stool weight and colonic motility were governed by the physical properties of the bran, such as particle size and water-holding capacity. Coarse bran was more than twice as effective both in holding water and in decreasing colonic pressure than fine-milled bran.

Because dietary fiber increases stool bulk by absorbing water, it is important to eat vegetables that have good water-holding properties. In addition to their value as sources of vitamins, the green and yellow vegetables recommended by Grandma many years ago are also included in the roughage group. To get a comparison of the water-holding capacity of various fiber-containing foods, I compiled data from several sources and listed them in Table 5. It is worth noting that there is considerable variation between the values obtained by different analysts on the same food product. This variability in results can be caused by several factors: not all labs analyze the same sample; some analyze fresh-weight foods, others extract the food pulps before analysis, and in some cases the foods were freeze-dried before testing, which resulted in enormous water uptake. Most foods we eat are fresh rather than freeze-dried, so fresh weights provide more reliable indices of water uptake.

Table 5 shows that fresh fruits such as apples, oranges, and bananas have a higher water-holding capacity as extracted pulp or pomace than as fresh fruit (whole apples, 17–46 vs. apple pomace, 235–509; fresh oranges, 20–56 vs. orange pulp, 176 grams of water per 100 grams of food). Also, the cereal brans have excellent water-holding capacity, a fact

Table 5. Water-Holding Capacity of Some Fiber-Containing Foods

Fiber Source	*Water-Holding Capacity*[a]
Sugar beet pulp	1449
Apple pomace	235–509
Apple, whole fruit	17–46
All Bran	436
Wheat bran	109–290
Rice bran	131
Oat bran	66
Corn bran	34
Cauliflower	28
Lettuce	36
Carrot	33–67
Orange, whole fruit	20–56
Orange pulp	176
Onion, whole vegetable	14
Banana, whole fruit	56
Potato, minus skin	22

[a]All values are in grams of water per 100 grams of food. Source: Gormley, 1977; Heller and Hackler, 1977 (their values are reported as "corrected" for fresh weight basis); and Chen et al., 1984.

that explains the popularity of high-fiber bran cereals as promoters of bowel regularity.

Some of the research on water-holding food fibers has employed human volunteers to study the effects on colonic function. In 1979, Stephen and Cummings examined the effects of 17 fibers contained in foods such as carrot, cabbage, apple, and potato and food additives and thickeners such as celluloses, food gums, carrageenans, and pectin. In general, gel-forming products, including celluloses, food gums, carrageenans, and pectin, held more water than the usual fruit and vegetable food fibers. Food gums are generally not present in significant enough quantities in foods to be considered useful for any health benefits. To obtain sufficient quantities to affect health, Cummings reported in 1978 that about 20 grams per day (about 3/4 ounce) of concentrated fiber from carrot, cabbage, apple, wheat bran, and guar gum added to human diets could increase fecal weights: 127% on wheat bran, 69% on cabbage, 59% on carrot, 40% on apple, and 20% on guar gum. Transit times through the GI tract were also shortened because of the increased bulk and softer stools. Consumption of unmilled cereals may be the most effective way of increasing fecal bulk because of the higher content of hemicelluloses in cereals, compared to fruits and vegetables.

Stephen and Cummings found that pectin held the most water but produced the smallest change in fecal weight, and wheat bran held the least water but increased fecal weight most. So they decided that dietary fiber does not exert its effects on fecal weight simply by holding water.

Some questions have also been raised against the water-holding hypothesis because hemicelluloses are partially broken down by enzymes in the GI tract and by bacteria in the colon. If the fibers are digested, then water-holding properties alone cannot explain the increased bulk and solid matter of the stools. Other research on colon bacterial activity has answered some of these questions.

COLON BACTERIAL ACTIVITY

Despite the original definitions of dietary fiber as plant matter that passes through the GI tract without being digested, we now have ample evidence that some forms of fiber are partially degraded. Cellulose and lignin resist colon digestion, but hemicelluloses can be partially broken down by GI tract enzymes (trypsin and pepsin) and by microbial enzymes (hemicellulases) in the colon. Also, some evidence suggests that pectins, gums, and mucilage can be partially degraded in the colon. In general, cereal bran fibers high in lignin, cellulose, and hemicellulose resist digestion better than fibers from fruits and vegetables, which are higher in pectin, gums, mucilages, and hemicelluloses. For example, in one study in which human volunteers were fed cabbage and wheat bran fibers, only 10% of the cabbage fiber was recovered in the feces, compared to 60% of the bran fiber. This result illustrated that both forms of fiber undergo some digestion, one more than the other, and suggested that increased ability to hold water was not the only effect of fiber in the colon.

This finding stimulated research on the activity of the microorganisms that live in our colons. Bacteria contain about 80% water, and they resist losing it despite the strong dehydrating forces of the absorption processes taking place in the left side of the colon. After cooking, plant materials are not as resistant to these absorption forces as are the bacteria. As bacteria "feed" on the fiber-containing materials entering the large intestine, they multiply and can comprise up to 50% of the fecal contents. As they break down the fiber, they produce some volatile fatty acids; some gases such as methane, hydrogen, and carbon dioxide; and some small sugars. The sugars and volatile fatty acids can serve as energy sources and can be absorbed into the body; the rest are excreted.

Dr. Abigail Salyers, who began investigating the digestion of plant cell wall fibers by human colon bacteria as a graduate student at Virginia Polytechnic Institute in the late 1970s, is still conducting active research in this area at the University of Illinois. She has described the effects of colon bacteria on various types of dietary fiber in an effort to understand their effects on fecal composition. There are 400–500 species of microorganisms in the large intestine, but just 100 of them account for 94% of the total. Many require some kind of carbohydrate for growth, mostly complex carbohydrates. Some species can use nonstarch carbohydrates, such as hemicelluloses, guar gum, dextrans, alginates, and pectin. These bacteria are anaerobic (that is, they function in the absence of oxygen) and ferment a wide range of fibers to different degrees. Dr. Salyers found that bacteria could account for 20–30 percent of the contents in the colon. She concluded that colon bacteria have developed a variety of enzyme systems for degrading dietary fibers. The amount of enzyme produced will probably be related to the concentration of the fiber present and to the bacterial growth rates.

GAS PRODUCTION IN THE GI TRACT

Gas produced by intestinal bacteria can be a problem when it causes abdominal bloating or pain. Some people seem to have more than normal amounts of gas in the large intestine. Unlike belching, which is caused by swallowing large amounts of air, rectal gas is produced by bacterial fermentation of undigested complex carbohydrates in the colon, by digestion of milk sugar (lactose) in lactose-intolerant people, by improper digestion of foods by those having various stomach or intestinal diseases, or by excessive consumption of certain legumes that are not fully absorbed. It is important to remember, however, that gas problems, no matter how great or how much pain is produced, are generally not fatal and are rarely caused by serious diseases. (Stress caused by serious illness usually is more of a cause for intestinal gas than is the food that was eaten.) Sometimes muscle spasms cause gas to get trapped in an intestinal loop, which produces bloated feelings and discomfort.

Cereal bran, very high in plant cell wall fiber, is now known to be partially degraded in the colon and might be expected to increase colon gas production when introduced into formerly low-fiber diets. This effect has been demonstrated in human studies where increased gas and bloating were noted shortly after bran was added to the diet. But improvement in bowel habits overcame the initial discomfort of the gas. Bond and Leavitt conducted extensive research on gas production and transit time, using various fibers. They concluded that bran serves as a relatively poor substrate for gas production by colon bacteria, so that abdominal symptoms reported by some subjects upon adding bran to their diets were unlikely to be caused solely by excessive gas. They showed that conditions such as diverticular pockets, which interfered with intestinal motility and with the passage of gas through the colon, caused the discomfort. Gas was well tolerated by normal subjects. Reports of excessive gas following ingestion of bran may also result from the fiber stimulating motility and increasing stool bulkiness in sensitive subjects resulting in increased gas production or the feeling of colon pressure.

♦

So, it is clear that appreciable fiber degradation does occur in the human colon. Fibers that are degraded include hemicellulose, guar, tragacanth, arabic and ghatti gums, pectin, and alginates. The simple sugars and volatile fatty acids that are not absorbed or metabolized further as energy sources go through the colon, to be stored until they are excreted. Sugars can also hold a certain amount of water, adding to the water content of the stored fecal matter.

To summarize: Dietary fiber's beneficial effects on the prevention of constipation and diverticular disease cannot be explained by its ability to hold water alone, because other factors can also increase stool weight. Plant fibers, such as cellulose and lignin, resist microbial degradation but they retain water, which increases stool bulk. Those fibers that are degraded increase bacterial growth. The bacteria are 80% water and add to the bulk and volume of the feces. Substances that increase bulk decrease transit times through the colon and increase motility. These effects, in turn, improve conditions for further bacterial growth, reduce water absorption in the colon, and result in softer stools. Also, some fibers act as gels that trap some of the gas bubbles produced in the colon, and trapped gas bubbles add further to the bulk. Thus, even though fiber is partially fermented in the large intestine, it still plays an important role by maintaining normal digestive processes and preventing constipation and its associated disorders.

Some people can consume any kind of fiber with no ill effects, whereas others experience cramps, stomach bloating, and excessive gas. For such people who wish to increase their fiber intake, it would be advisable to first introduce small amounts of the fiber into the daily diet for a period of 2 or 3 weeks until the GI tract adjusts physiologically to the new fiber. After discomfort subsides, larger amounts of the fiber can be added to the diets of most individuals without distress.

6

Can Fiber Cure Cancer, Diabetes, and Other Serious Diseases?

If you've read the boxes of some of the high-fiber breakfast cereals, you've seen statements like these:

> "The National Cancer Institute believes eating the right kinds of foods may reduce your risk of some kinds of cancer . . . EAT HIGH-FIBER FOODS . . ."
>
> "Following the National Cancer Institute's high-fiber, low-fat dietary recommendations to reduce your risk of some kinds of cancer is just a matter of eating less of some kinds of foods and more of others . . . eating 20–30 grams of fiber every day . . ."
>
> "Growing evidence suggests that food fiber (often called roughage) may have a more important role in our diets than previously thought . . ."
>
> " . . . Many health professionals agree that it is important for everyone to include fiber in their diet . . ."
>
> "Did you know . . . dietary fiber is currently linked to healthier digestion as well as the reduction of risks of diabetes, hypertension and heart disease, obesity, and some forms of cancer . . ."
>
> "Eat Healthy and Slim Down. The HIGH-FIBER DIET PLAN THAT WORKS . . ."

These statements are then followed by declarations about why their brand of breakfast cereal can help you avoid these dreaded diseases and live a healthier life.

One of today's most feared words is CANCER. Many people, especially those in middle age, almost panic when they hear it. When they read their cereal boxes, they do not see the words "risk of certain diseases", they only see "cancer". But there are many forms of cancer, and fiber is not a cure-all. As we have seen, Dr. Denis Burkitt has advocated a high-fiber diet to help people avoid such diseases as appendicitis, varicose veins, diabetes, and colon cancer.

We've already discussed the chemical properties of foods that can influence gastrointestinal disorders. They are believed to be related to the increased water-holding capacity of food fibers. Fiber's ability to hold water softens stools, speeds up their transit time through the GI tract, and binds bile acids. Softer stools mean less pressure is required to eliminate wastes, thus relieving constipation, diverticular disease, and hemorrhoids. Now let's turn our attention to fiber's reported effects on colon–rectal cancer, diabetes, and other diseases.

COLON–RECTAL CANCER

Advancing age does not change the body's requirements for regular intakes of the basic nutrients, only the amounts. For example, less fat and protein are required because the body needs fewer calories to sustain decreased physical activity and reduced synthesis of body proteins. But the need for complex carbohydrates may increase. Dr. Burkitt feels that the drop in consumption of coarse cereal grains and increased consumption of animal fats three to four decades ago in America and Britain are responsible for the increase in colon–rectal cancer. It should be stressed, however, that there is still no unequivocal evidence that a high-fiber diet helps prevent colon–rectal cancer. But there is enough scientific evidence to suggest that high-fiber diets help to prevent the onset of this cancer. Because of the complex nature of the different fibers, some of these studies have yielded conflicting results. One major deficiency is the present lack of more definitive research on the link between the disease and low-fiber diets as well as with diets high in fat.

One point that does seem recurrent in the reported studies on colon cancer in humans is that colon–rectal cancer appears to be more closely related to economic status and modern Westen living habits than do other forms of cancer. Most scientists and physicians agree that diet may be the most important cause, though there are still differences of opinion as to *which* dietary factor is the primary cause. Is it increased fat intake or is it decreased fiber intake? Or is it both? The answer is not a simple "either–or", but it is very likely a matter of "both–and". The manner in which cancer-causing compounds are formed in the colon is not known, but some researchers believe that their carcinogenicity can be increased if they are concentrated in the fecal matter in the large intestine and remain in contact with the inner wall of the colon for long periods of time. This inner wall of the colon is where cancers begin to form.

Although there is no *absolute* proof that dietary fiber will indeed prevent colon–rectal cancer, there is enough growing evidence that an adequate fiber intake can greatly reduce the *risk* of cancer, but the exact mechanism of this process is still being debated. Some researchers feel that fiber's role in binding and removing bile acids in the feces is a factor. Some feel that fiber functions by speeding the movement of waste materials through the colon, thereby reducing contact of the carcinogens with the colon wall. Others feel that some fibers, e.g., pectins, gums, and hydrocolloids, that hold large quantities of water to form gels move through the colon and engulf the carcinogens in a way that reduces their potential danger by removing them in the feces. Still others feel that some types of fiber act as antioxidants that "defuse" the carcinogens, and some feel that fibers bind directly to mutagens (materials that can cause healthy cells to mutate into cancer cells) and remove them from the GI tract.

Dr. Burkitt believes that the extra volume of stools formed in high-fiber diets decreases the relative concentration of these carcinogenic compounds, and the faster movement of fecal materials through the intestinal tract shortens the time that the intestinal walls are exposed to these toxins. The decreased exposure to these substances and their faster excretion in the feces are believed to be the reason for the lower incidence of colon cancer in the rural African populations he studied. But it is worth emphasizing here that diet is not the sole difference between rural populations that consume more high fiber and coarse grains and Western populations that consume low-fiber diets. Western populations are generally more sedentary; the rural populations Dr. Burkitt studied are natives of a developing country who live an entirely different life style with fewer or no conveniences, more physical activity, and less fat in the diet.

Hugh Trowell, an associate of Denis Burkitt, wrote in 1976 that epidemiological data have suggested a negative correlation between age-adjusted annual death rates from colon cancer and cereal intakes in 38 countries and with crude fiber intakes. He said that certain nuclear dehydrogenating *Clostridia* (colon bacteria) can produce carcinogens from bile acids. But in Uganda there is an extremely low prevalence of colon cancer. The primary staple of the diet there is plantain, a bananalike starchy fruit that is very high in hemicelluloses that are degraded in the colon. This degradation produces acidic feces that are rarely found to contain the *Clostridium* strain that is the important dehydrogenating species. In other words, populations that consume large amounts of high-fiber foods and unrefined cereals have relatively fewer cases of colon–rectal cancer than Western populations that consume less fiber and more animal protein products and more fat.

Thus far, the evidence has been acquired by epidemiological studies of different populations with different eating habits and different life styles. These studies have also suggested that some fruits and vegetables contain potential cancer preventing compounds such as carotenoids, phenolic acids, flavonoids, and lignins. Many of these compounds derived from plants have applications in the food, drug, and cosmetic industries because they are GRAS (generally recognized as safe) and have desirable chemical and/or biological properties. These "anti-cancer" foods are sometimes referred to as "designer" foods because of the following reasons:

- they are acceptable and are naturally concentrated in or supplemented by these epidemiologically important plant chemicals from fruits and vegetables, or
- they have been proven to be anticarcinogenic in animal tests, or
- they have unique structures and/or metabolism that may be used as markers in compliance tests.

The American Cancer Society guidelines recommend increasing the intake of high-fiber foods, including those vegetables from the mustard family—broccoli, cabbage, cauliflower, brussels sprouts, mustard greens, and turnip greens) and vegetables having high concentrations of vitamins A and C. They also recommend reducing total fat intake.

More recently, in the September 6, 1989 issue of the *Journal of the National Cancer Institute*, Jerome DeCosse and his co-workers published the results of their 4-year study on the effects of cereal fiber on precancerous polyps in the colon. They reported that a diet high in fiber appears to promote shrinkage of the polyps, which should lower the risk of colon–rectal cancer in humans. Previous studies had suggested that dietary fiber can reduce the risk of cancer, but this new study provided the first evidence of a direct effect in humans—and these results were achieved with a wheat bran cereal high in insoluble fiber, a food available in all grocery stores. This result suggests that breakfast cereals high in bran can reverse the normal progression of potential colon cancers by inhibiting the precancerous lesions (polyps).

Several years ago the U.S. government sponsored three major food-related surveys: the Ten-State Nutrition Survey (1968–1970), the First Health and Nutrition Examination Survey (1971–1974), and the Nationwide Food Consumption Survey (1977–1978), referred to as Ten-State, HANES-I, and NFCS, respectively. These surveys, in general, confirmed what Dr. Burkitt had theorized: that Americans have generally begun to eat less food by volume, but have been consuming more calories from the smaller amounts of food, which suggests that Western diets have become higher in fat and lower in fiber. The surveys collected data on all age groups and uncovered the striking fact that it was the older age groups that tended to consume the increased-calorie diets, contrary to common belief that the younger age groups eating fast food products were the major reason for the national diet being higher in fat. Those surveyed consumed more mixed-protein meals (such as beef and cheese,

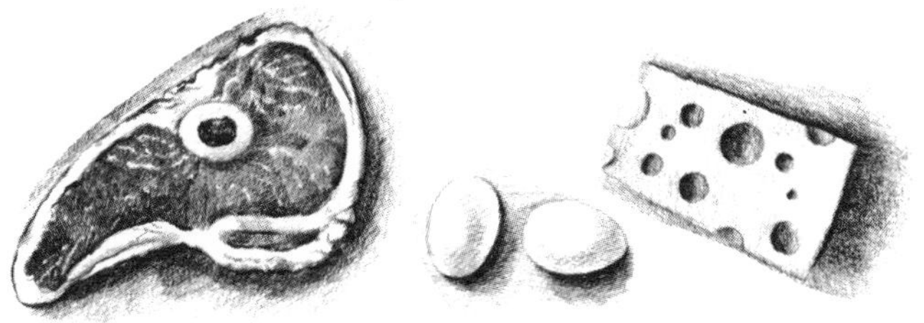

ham and cheese, bacon and eggs, or beans and meat), poultry, vegetables, fruits, red meat, sauces and gravies, and fewer dairy products and nonsugar diet drinks. Cereal grain foods seemed to remain stable overall, but the older age groups showed a slight gain in consumption possibly due to increasing interest in bran and its beneficial effects.

As with most serious diseases, some people are more susceptible to cancer than others because of certain risk factors. For colon–rectal cancer, these are

1. A family history of colon–rectal cancer or polyps.
2. Increased age.
3. Life style. The sedentary city dwellers and those exposed to pollution are more susceptible than people who are physically active, who live in rural areas, and who are exposed to less pollution.
4. Diet and excessive weight.

Diet and weight are the easiest of these factors to control; you can't change your ancestors or your age. The first dietary component we hear a lot about is fat. What about high-fat diets and cancer? It is not possible to conduct experiments on fiber and cancer prevention in humans, so this research has been conducted on small animals. Dr. Kent Erickson has conducted research on nutrition and cancer since 1975. He reported in 1984 that there is now sufficient evidence that dietary fat may be an important factor contributing to the development of cancer, especially cancer of the colon, breast, skin, and prostate. Saturated fat is the major culprit—not only animal fat, but hydrogenated vegetable fats and shortenings as well. Some scientists have implicated linoleic acid as the critical fatty acid in tumor formation because of its tendency to oxidize in the body and produce free radicals—compounds that can initiate reactions leading to cancer development.

Dr. David Kritchevsky has also investigated the role of fats in cancer development. In 1987 he reported that fat alone was not the culprit. High-fat diets are more efficient promotors of tumor growth than are low-fat diets, but only unsaturated fatty acids are more efficient tumor promotors than saturated fat. The PUFA (polyunsaturated fatty acid) effect is due in part to the essential fatty acid, linoleic acid. The general conclusion from various laboratories is that replacing saturated fats with PUFA, such as corn oil, is not necessarily beneficial because PUFA can enhance tumor growth more than saturated fats. Although high-fat diets are definitely better promotors of tumor growth than low-fat diets, too many calories in general—from *any* foods—are just as bad as high-fat diets. Dr. Kritchevsky believes that the level of calories in the diet rather than the level of fat consumption may be the most important dietary factor in preventing cancer development (although not in curing it once it is well established). Regardless, an increased intake of fiber can help by decreasing the metabolism and absorption of fat (*see* Chapter 2). One sure way to avoid excessive intake of fat and calories is to substitute complex carbohydrates and fiber for fatty foods. Doing so would replace fats (measured at 9 calories per gram) with starches (at 4 calories per gram) or with "no calorie" dietary fiber.

One of the most extensive studies conducted on humans was done by Dr. Roland Phillips and his staff at Loma Linda University, California (Phillips and Snowden, 1986). Their report summarized 21 years of studying 25,000 California members of the Seventh Day Adventist Church. True Adventists prohibit habits that can adversely affect health, such as smoking and drinking of alcoholic beverages. They also recom-

"You can dye your steak blue if you want to, but it's still considered 'red' meat."

Reprinted with special permission of Cowles Syndicate, Inc.

mend (but do not require) that members avoid animal foods, caffeine-containing beverages, and hot, spicy foods. About one-fourth of the Adventists are vegetarians who also eat dairy products and eggs, a fourth are strict vegetarians, and half follow a nonvegetarian diet.

From 1960 to 1980, 25,000 California Adventists age 35 and over were followed carefully. In addition to the very low incidence of lung cancer (doubtless related to their nonsmoking tenets), these people also suffered less from other forms of cancer. Complete vegetarians had a marked reduction in the risk of fatal colon cancer, coronary disease, stroke, and diabetes compared to Adventists who consumed meat heavily. Among Adventist men who consumed few or no animal proteins, the risk of fatal prostate cancer was a third of that for Adventist males who consumed animal proteins regularly. For cancer of the colon, breast, and prostate, as well as diabetes, the risk among Adventists was substantially below that of the general population, a finding that strongly suggests that dietary habits (plant vs. animal proteins) may account for the low risk. (The only cause of death in which Adventists equaled the general population was traffic accidents.)

Although Adventists consuming animal proteins showed greater risk for cancer than did Adventists consuming only vegetable proteins, Dr. Phillips did not conclude that meat use was unequivocally related to fatal cancers of the colon and breast. Because he did not do an itemized study of the specific foods eaten by the populations under investigation, he could not ascribe the difference to the red meats and animal fats in diets of one group or to lowered fat in the all-vegetable diets. However, the fiber content of all vegetable diets is notably higher than that in animal protein diets (which presumably contain fewer vegetables). The proteins themselves also differ chemically between plant and animal sources. Dr. Phillips did conclude that consumption of animal proteins was strongly related to the risk of fatal prostate cancer and fatal coronary disease in males, and fatal diabetes in both males and females. The evidence acquired through this extensive study may not have provided unequivocal proof, but it strongly suggests that dietary protein, like fat and total calories, can be a factor in cancer.

Little research has been done on the effects of exercise on cancer, but the American Institute for Cancer Research newsletter reported in the summer of 1987 that colon cancer risks are greater in people at all income levels and all racial groups whose jobs require little physical activity. Some studies have showed that women athletes have lower risks of breast and reproductive-tract cancers than nonathletic women. Remember, however, that athletes and those who exercise regularly are more likely to take care of their bodies, control their weight and monitor the kinds and amounts of food they eat, and not smoke or drink alcohol to excess. Thus it is difficult to ascribe any differences in cancer rates between athletes and nonathletes to exercise alone. Exercise is beneficial, but it appears that the combination of exercise, weight control, and diet may be responsible for the change noted. If any recommendation is possible after review of the published literature it would be for sedentary, nonexercising people, especially those over 50, to keep an open mind and a closed refrigerator. Sample the bowl of fruit instead.

Kies and co-workers in 1981 offered another mechanism for the role of fiber in cancer prevention. They suggested that fiber aids in the removal of irritants and carcinogens from the GI tract by binding to lithocholic acid within the intestine. Lithocholic acid is a potentially toxic bile acid when degraded. When it is unbound it may be a factor in causing gallstones and gallbladder disease, but with a high-fiber diet, it binds to the fiber and is excreted via the feces. Others at the University of Georgia (Noorman et al., 1983) and the University of Lund in Sweden (Sjodin et al., 1985) showed that dietary fibers, such as wheat bran, carrot, guar gum, pectin, corn bran, cellulose, barley, sorghum, sugar beet pulp, and oat bran, bind to some compounds formed in fried meats. The compounds may be associated with intestinal diseases, including colon cancer. Water-soluble fiber components were responsible for most of the binding. If these toxic materials are bound and excreted, their contact with the intestinal walls should be minimized.

DIABETES

Almost 35,000 Americans die from diabetes and its complications each year. Because diabetes hastens atherosclerosis, patients with diabetes die from coronary disease two to three times more frequently than nondiabetics, Steiner reported in 1981. Traditionally, most cases of diabetes have been treated with insulin or drug therapy, with minimal mention of diet therapy. About 10% of the diabetic population has type I (insulin-dependent) diabetes mellitus, which frequently develops rapidly during childhood. Type II (non-insulin-dependent) diabetes mellitus occurs more frequently in overweight older people.

Anderson and Tietyen-Clark reported in 1987 on their studies of the effects of high-fiber, low-fat diets on diabetes and found that high fiber intakes consistently improved control of glucose levels in the blood, reduced insulin requirements, lowered serum levels, and promoted weight loss and maintenance. They fed diabetic patients high-carbohydrate, high-fiber diets of fruits, oats, oat bran, barley, and legumes (especially effective were oat bran and dried beans) that provided 70% of the calories as carbohydrates and 70 grams of fiber daily. Average fasting serum glucose dropped from 186 to 156 mg/dL (milligrams of glucose per deciliter of serum) for type I diabetics and from 161 to 135 mg/dL for type II, despite significantly lower or discontinued insulin doses. Insulin doses were decreased 38% for type I and 95% for type II diabetics. Anderson and Tietyen-Clark reviewed nine studies of high-carbohydrate, high-fiber diet therapy for diabetics in which average insulin dosage decreased by 40%. They concluded that high carbohydrate, high-fiber diets combined with regular exercise and healthy life style can significantly reduce the incidence and severity of diabetes in the United States.

The precise mechanism for the sparing action of fiber on insulin requirements is not known. It is well known that insulin is involved in the metabolism of sugar. Insulin functions by helping sugar to get through the cell membrane and into the cell where it can be metabolized. The primary tissues where sugar metabolism fails in diabetes are the muscles and connective tissues, especially the fat-synthesizing cells. When a person eats carbohydrate, enzymes break it down to sugars that get into the bloodstream and raise the serum glucose level, signaling the pancreas to secrete insulin. The more sugar, the more insulin that is secreted. Insulin is also transported through the blood to the cells where it attaches to a specific point on the cell's surface and allows the sugar to enter the cell.

Dr. Phyllis Crapo has recently shown that all starches do not liberate sugar at the same rate. Some take longer to break down and be absorbed, which results in a slower, more manageable rise in blood sugar and insulin. These discoveries are especially important for diabetics. Dr. Crapo fed human volunteers potatoes and rice and found dramatic differences in their metabolism. Rice gave a flat glucose response, but potatoes gave a rapid response, almost as if the volunteers had consumed glucose directly. Then she fed test meals of corn, bread, rice, and potatoes—the four major starches—to diabetics and to older people with impaired glucose tolerance. Blood sugar (glucose) rose least in response to rice, more with bread and corn, and most with potatoes.

Blood sugar rose only half as much in response to legumes as it did to cereal starches. Dairy products, including ice cream, generally made blood sugar rise slowly, but cheese–bread combinations unexpectedly resulted in a rapid rise, similar to that of bread alone. Bread and beans combined, however, gave slow rises more characteristic of the beans than the bread (great news for those who enjoy the tasty New Orleans dish, red beans and rice with French bread). Dr. Crapo and her team also found that pasta products were lower in their blood sugar effects than cereals. However, for nondiabetics, whole wheat products may still be more advisable because of the beneficial effects on the colon. Dr. Crapo's work is important for diabetics who must avoid rapid rises in their blood sugar.

Some medical researchers now feel that the reason some people develop diabetes as adults may be because they eat too much of the complex carbohydrates that give rapid rises in blood glucose. In developing countries, people eat diets rich in coarse starches, cereal, and beans that cause slow blood glucose rises; hence they tend to have little diabetes or colon and heart diseases.

Some diabetics produce no insulin. Others produce such small amounts that it is absolutely necessary to get maximum use of the insulin they do have by controlling the amounts and the rates of sugar entry into the blood. By careful control of the composition of their diets, emphasizing primarily complex carbohydrates rather than sugars or starches that release sugar into the blood rapidly, diabetics can prevent sudden surges of blood sugar and demands for extra insulin. Diet may also increase the cell's sensitivity to insulin so that diabetics get more use of the insulin they do produce and, therefore, need less or no insulin therapy. A diabetic diet containing ample protein and vitamins, that is high in complex carbohydrates and fiber, low in sugar and fat, should be nutritionally sound for this purpose and should be conducive to better control and prevention of obesity.

These results have been confirmed recently by other researchers. In 1985, Anderson fed high pectin, guar, or wheat bran diets to 25 type I diabetics for 3 weeks. Insulin needs decreased by 38%, and fasting blood glucose values decreased by 16%. In Sweden in 1984, Karlstrom and his co-workers fed 14 type II diabetics a diet containing either 18.9 grams or 42.4 grams of dietary fiber in a randomized order. The mean glucose level and urinary glucose excreted were significantly lower in patients consuming the higher fiber diet, though serum insulin levels were similar in both groups.

Oats have also been beneficial for diabetics. In 1985, Hardin showed that diabetic men who consumed 100 grams of oat bran daily ceased needing insulin shots to control their blood glucose. This result was attributed to the high content of water-soluble gums in the oat bran. However, a half-cup of cooked oats as oatmeal weighs about 30 grams, so, to ingest 100 grams per day would require larger portions of oatmeal plus other products such as oatmeal cookies and oat bran muffins.

The insulin-to-glucose ratio was 33% higher in patients on the higher fiber diet. It appears probable that a high-fiber diet can be more beneficial for diabetics than for nondiabetics by lowering insulin requirements and improving control of serum glucose levels. It should be emphasized, however, that diabetics should not switch over to a high-fiber diet without careful monitoring by a physician until the proper diet intake and composition are established.

OTHER DISEASES

Atherosclerosis has already been discussed in detail in Chapter 3. Obesity is discussed in detail in Chapter 4.

Fiber may also affect recurrence of peptic ulcers in humans. Rydning and Berstad reported in 1986 on a 6-month follow-up study of 73 patients with recently healed duodenal ulcers. They found that the recurrence of ulcers was significantly higher in the group fed a bland, low-fiber diet; that is, 80% vs. 45% in a group of patients who ate a high-fiber diet. Patients on the low-fiber diet also tended to relapse more frequently than patients on the high-fiber diet. It appears that the traditional ulcer diet of small, bland meals to reduce secretions of acid and pepsin and to reduce gastric motor activity may not be as useful as a high-starch and high-fiber diet in peptic ulcer treatment.

◆

It should be obvious now that dietary fiber is not a miraculous medicine that will cure our most feared diseases. But it can influence several processes after the ingestion of food. In the mouth, fiber stimulates chewing, causing saliva and the enzyme amylase to start flowing. It helps clean the teeth of food particles and acids produced there, which should help reduce the formation of dental plaque, a major carrier of gum disease and tooth loss. Then, for the obese or those wishing to control weight, it can produce a feeling of satiety and reduce caloric intake. Dietary fiber can also delay the rate at which the stomach empties its contents, because solids are more viscous and, therefore, are emptied more slowly than liquids. In the intestine, the fiber passes intact until it reaches the colon. There it can be broken down by bacteria and used as an energy source, to hold water, and to add fecal bulk; processes described earlier.

The best recommendations for a prudent diet that can be beneficial in the prevention of these diseases are those made earlier: eat a balanced diet, avoid too much fat–especially saturated fats and cholesterol–eat adequate complex carbohydrates and fiber-containing foods; avoid excess sugar, salt, alcoholic drinks, and smoking; and exercise regularly to avoid overweight and obesity.

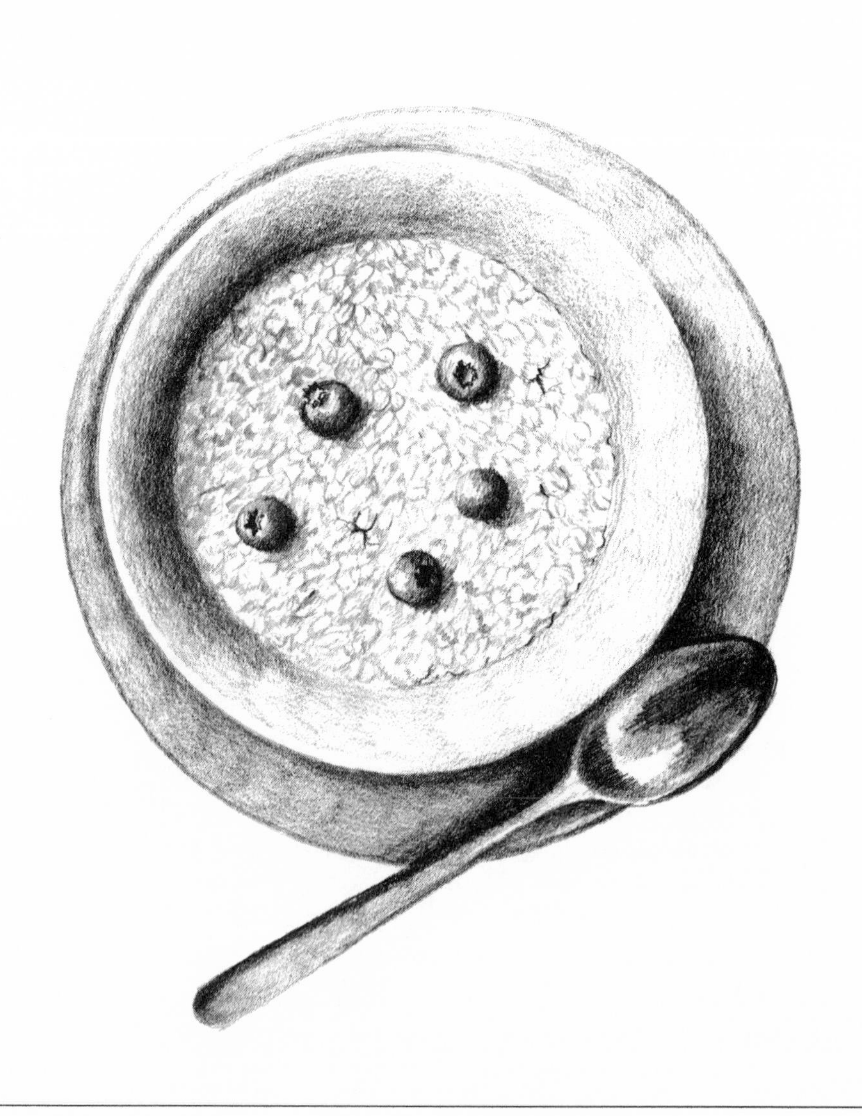

7

Fiber and Mineral Nutrition

The bran layers of cereals are not just major sources of dietary fiber. They also contain most of the minerals present in the cereals. Why, then, is there such concern about the loss of minerals from a diet high in fiber?

With a *balanced* diet, there is apparently no danger of normal, healthy individuals developing a deficiency of trace minerals. However, with fad diets or diets containing excessively high amounts of fiber, it is possible to lower the bioavailability of trace minerals, and that may lead to deficiencies of iron, copper, zinc, calcium, and magnesium.

The key word here is "bioavailability". The presence of any mineral in a food or food product does not necessarily mean that all of the mineral will be "available" to the host or be absorbed by the body. If the minerals are bound to some insoluble material from which they cannot be separated in the digestive tract, they will pass through it and be excreted in the feces or the urine.

Wheat, rice, and oat brans are good sources of minerals such as iron, calcium, zinc, phosphorus, and magnesium. But those brans also contain major deposits of phytic acid, an organic compound that contains excellent binding sites for the metals. Calcium, magnesium, and zinc are often bound to phytic acid in cereals. Many scientists feel that phytic acid lowers the bioavailability of these minerals from cereals (and also dry beans, another good source of phytic acid). Thus, when unrefined cereals are consumed, calcium, magnesium, and zinc are already bound to the phytic acid. There is no absolute proof, however, that the minerals are released from the phytic acid and that it binds other minerals in place of those it releases. In addition, the leavening process of baking produces enzymes that can reduce phytic acid. And certain microorganisms of the large intestine produce enzymes that can release the minerals for possible absorption.

Excessively high fiber diets may affect mineral absorption by certain individuals, but the results may not be due solely to fiber, but to phytic acid. The phytic acid controversy arose about a decade ago when a population of Iranian males was found to be suffering from zinc deficiency and to have undeveloped sexual characteristics that rendered them sterile. Their diet contained high amounts of unleavened high-bulk, whole-grain black bread and beans—with little or no animal protein. Animal protein is high in zinc; when it was added to their diet, the sterility was reversed, and blood levels of zinc, calcium, and magnesium were elevated. (It was not fully clear whether the sterility was due to deficiencies of zinc, protein, or both.) As will be seen later, one type of interaction between phytic acid and fiber affects the binding of minerals. This binding is also influenced by protein.

The bioavailability of minerals from high-fiber diets is, like the fibers themselves, a very complex matter. There are too many published reports of such research to review here; only a few will be mentioned to illustrate the interactions between fiber and minerals. As usual with such a proliferation of reports, some seem to conflict with findings and conclusions of others, especially in the controversy on fiber or phytic acid as the culprit.

Minerals have various functions in the body, all of them important. Iron is required for proper synthesis of hemoglobin, prevention of anemia, and as a cofactor of certain enzymes in the body. Copper is present in certain enzymes and is required for proper utilization of iron in anemia prevention. Calcium is probably the major mineral in the body; it is required, along with phosphorus, for proper bone and tooth formation. Calcium also plays a role in muscle tone. Zinc is required for proper sexual development in males, for healing of wounds, maintenance of appetite and the sense of taste, and is present in some important enzymes. Magnesium is a cofactor with some enzymes; it is required for proper utilization and metabolism of carbohydrates, and is present in bone.

How does fiber interact with minerals? Because the various fibers have chemical structures that can "act acidic" by attracting metals, they can bind free minerals during digestion in the body. Unless the minerals are released in some way further down the GI tract for absorption by the body, they will be excreted in the feces. Fiber and phytic acid are not the only factors that affect bioavailability of minerals. Other factors to consider include

- the digestibility of the food containing the minerals,
- whether the body is saturated with a particular mineral and can store no additional amounts,
- interactions with other components in the food such as low solubility organic acids, (e.g., oxalic acid in spinach) or some inorganic acids (e.g., phosphoric acid–phosphate),
- the health of the individual, and
- whether the stomach and GI tract are too high or too low in acid.

For example, green leafy vegetables like spinach and cabbage are good sources of vitamins, minerals, and fiber. But they also contain oxalic acid, an organic acid that binds calcium to render it insoluble. If excessively large amounts are consumed without additional sources of calcium from extra milk or cheese, calcium oxalate will be formed and excreted via the kidneys. Calcium oxalate is an insoluble compound, so crystals sometimes form in the kidneys and cause pain and further problems not related to the fiber or phytic acid.

HEMICELLULOSES AND MINERALS

During our research on the dietary fiber of rice, we examined the binding of iron, copper, calcium, zinc, magnesium and manganese by the hemicelluloses of long- and medium-grain rices, their subsequent release by certain enzymes present in the GI tract (trypsin, pepsin, and hemicellulase), and the effects of phytic acid and protein on binding under laboratory conditions simulating those found in the intestines, and also by feeding to live animals. I've chosen to describe the research in some detail to give you an idea of how complicated it can be. We selected trypsin and pepsin, normal protein-digesting enzymes present in humans, because the structure of the hemicelluloses contains some protein portions. We also used hemicellulase, an enzyme not found in mammals but present in some microorganisms of the large intestine; it breaks the carbohydrate or sugar bonds of the hemicelluloses.

After isolating, purifying, and chemically characterizing hemicelluloses from both rice bran and milled rich endosperm, we investigated the binding of selected minerals to specific hemicelluloses under varying conditions of temperature, salinity, and pH (acidity–alkalinity), simulating those in the intestinal tract. We accomplished this by carefully weighing amounts of the hemicelluloses and minerals, placing them in the reaction mixture, and setting the reaction vessel in a bath at body temperature. Samples were removed at intervals up to 2 hours and transferred to dialysis bags. These bags have a permeability similar to the walls of the intestine; large molecules are retained in the bags, but free minerals pass through.

The apparatus used in this procedure may be seen on the opposite page. The dialysis bag contains fiber and minerals in the reaction mixture. It was placed in the long tube-funnel shown at the right in the figure, designed to hang in the large cylinder at the left in a position such that the bottom tip is 2–3 millimeters (about 1/4 inch) from the cylinder bottom. Distilled and deionized water was added to a depth of 2–3 millimeters above the rim of the funnel. The apparatus was then set in an incubator undisturbed for 48 hours to allow the unbound metal to diffuse through the dialysis membrane. As metals diffuse out, being heavier than water, they drop to the bottom, causing fresh water at the top to be displaced over the funnel rim. This keeps fresh water continuously on the surface of the dialysis bag. For treatment with enzymes, the reaction mixtures were allowed to stand with the added enzymes up to 16 hours more to simulate the extra time foods require to pass through the GI tract.

A typical series of tests on copper, zinc, and iron binding showed the effects of the enzymes. At a ratio of 10 milligrams of metal reacted with 1 gram of hemicellulose, the amounts of minerals bound are small

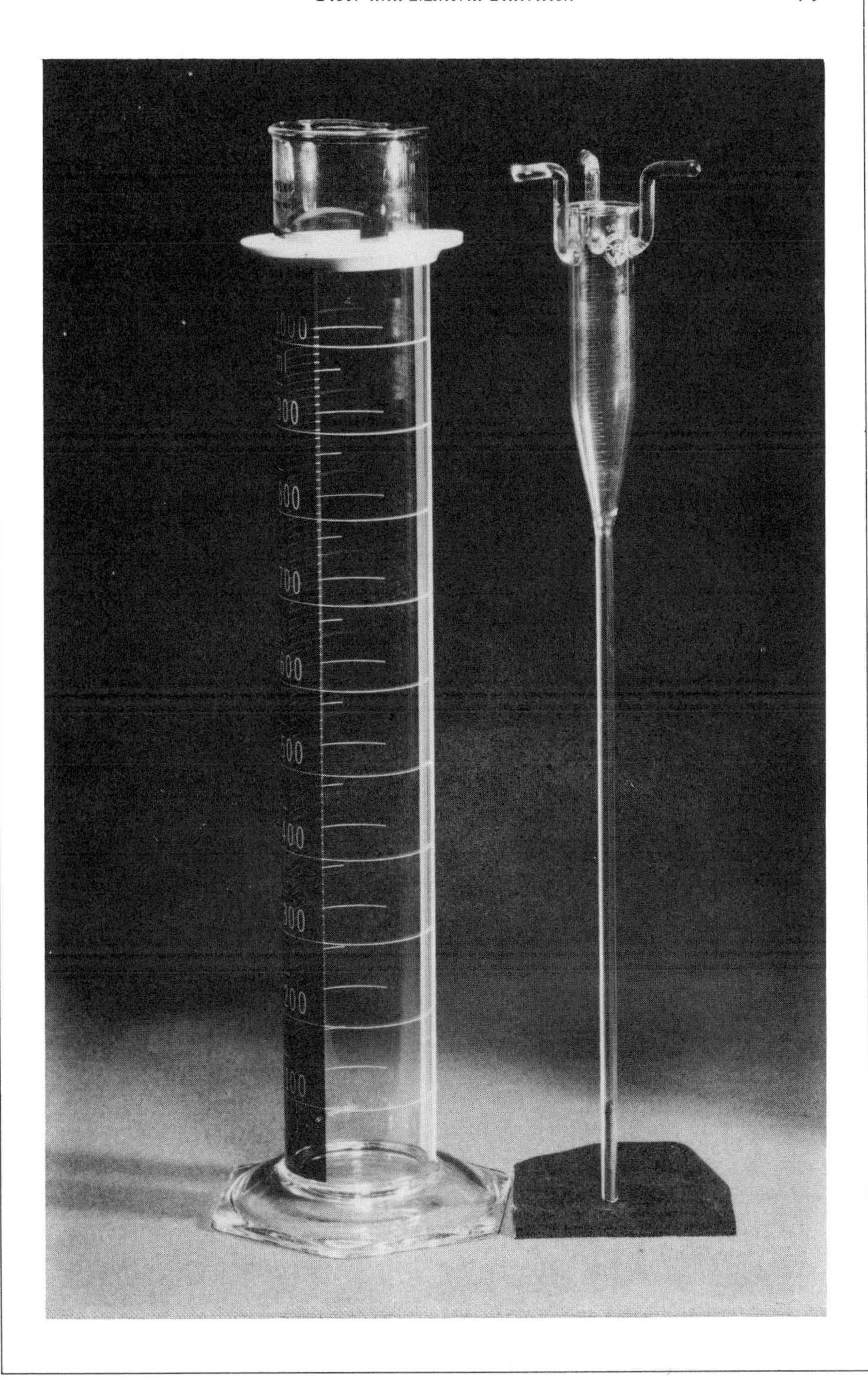

(less than 0.01%), but all three were bound in varying amounts to the hemicellulose. However, the enzymes are able to release some metals from the bound state, a result suggesting that, under normal conditions in the body, some bound mineral should be released by enzymes for absorption by the body. For example, trypsin releases 59% of the copper, but little or none of the zinc and iron from the insoluble hemicellulose. Trypsin releases less copper from water-soluble hemicellulose (41%), but releases more zinc (21%) and iron (37%). Pepsin, another enzyme that appears to hydrolyze protein portions of the hemicellulose, releases the metal in a manner similar to that of trypsin.

Hemicellulase, on the other hand, breaks the sugar linkages of the hemicelluloses and reacts differently than the protein-splitting enzymes. It releases more zinc (64%) than copper (26%), but no iron from the insoluble hemicellulose, and almost all of the zinc (92%), 21% of the copper, and 36% of the iron from the soluble hemicellulose. This result indicates that copper may bind primarily to the protein portion, and zinc to the carbohydrate portion of the insoluble hemicellulose in such a manner that enzymes in the GI tract can release large amounts. The bound iron is not released to any great extent. With the water-soluble rice hemicellulose, copper and zinc again seem to bind primarily to the protein and carbohydrate portions, respectively, but iron is apparently bound in a different manner. Almost two-thirds is released by pepsin, and one-third by either trypsin or hemicellulase. This finding indicates that minerals do bind to hemicelluloses, but enzymes of the GI tract can break some of the binding sites to release the minerals for potential absorption by the body.

I've mentioned that phytic acid and protein can affect the binding of minerals. Our work showed that protein in the diet can affect the binding of minerals to water-soluble rice hemicelluloses. As the amount of protein in the reaction mixture increased, the binding of copper, zinc, and iron was depressed by varying degrees, zinc more than the other minerals. This result emphasizes the importance of eating balanced meals and avoiding fad diets with excessively high fiber contents. A meal containing extra fiber should also contain adequate protein, not only to provide amino acids for the rebuilding of the body's proteins, enzymes, and hormones but also to assist in proper metabolism of essential minerals.

On the basis of these results of mineral binding to rice hemicelluloses and their possible release by enzymes, we proposed theoretical mechanisms for the binding of the minerals and their release by enzymes (Mod et al., 1981, 1982; Normand et al., 1987). Copper is apparently bound in the protein portion of the hemicellulose. When the protein-splitting enzymes destroy the configuration or structure that is holding the minerals, the copper is released for potential absorption. For zinc, the binding appears to be by sugar bonds in the hemicelluloses. As the enzyme cleaves sugar bonds, the sites binding the zinc are weakened so that zinc is released for possible absorption. Iron seems to be bound by both protein and sugar, so that both proposed mechanisms may function in the release of bound iron for potential absorption by the body.

Similar results were obtained by Dintzis and Watson in their 1984 study of iron binding of wheat bran fiber at human stomach pH. They noted that the binding of iron by all varieties of wheat brans examined was dependent upon time and concentration of iron in the solution. Iron binding could be altered greatly by various treatments that changed the conditions. They concluded that binding mechanisms other than phytic acid were involved. Camire and Clydesdale in 1982 found that additions of 60–180 milligrams of vitamin C with wheat bran samples could prevent the binding of iron to the wheat brans. (At least 50 milligrams per day is the recommended daily allowance). Again, all of this research seems to emphasize eating a balanced meal rather than a high-fiber fad diet.

To examine the effects of fiber on mineral binding in a living animal, we divided weanling rats into three groups of five rats each, in separate cages, for a 2-week study, 7 days for depletion and 7 days on special diets (Mod et al., 1985). One group was fed the control diet with no fiber added, one was fed the control diet plus 3% purified cellulose, and the other was fed the control diet plus 3% purified water-soluble rice bran hemicellulose. (The control diet consisted of protein, corn oil, corn starch, vitamins, and dextrose.) Whole brown rice was not used as the fiber source for fear of introducing a blood coagulation factor that we had found in uncooked rice in earlier research. Addition of such a factor could influence the results and lead to erroneous conclusions.

Both fibers fed at 3% levels caused slight increases in excretion of minerals in the feces. The purified cellulose retained more of the bound minerals than did the rice hemicellulose, resulting in greater excretion from the body. This effect could be especially important to persons trying to control or lose weight by consuming large amounts of the specialty breads that contain 5–10% cellulose. Weight loss can be achieved, but caution should be observed to avoid a mineral deficiency or imbalance. Our results do show that extra cellulose or hemicellulose fed continually in each meal, every day for an extended period, can affect mineral balances. This finding suggests that high-fiber fad diets should not be continued for long periods of time. In a balanced diet containing whole brown rice or white milled rice as the carbohydrate, such losses should not occur because the concentration of hemicellulose in whole grains is lower than 3%.

WHEAT BRANS

The most highly recognized and widely used cereal fiber is probably wheat bran because of its ready availability from milling of wheat to produce white flour and because of its low cost relative to other cereals. Because it has been used for many years in ready-to-eat wheat bran breakfast cereals, interest in the chemical properties of wheat bran has attracted a great deal of research attention. Though chemically related to rice hemicelluloses, the primary fiber of wheat bran is frequently called pentosan because of the high concentration of the five-carbon sugars xylose and arabinose. Wheat pentosans, like the hemicelluloses, have different chemical structures and binding properties, and they have also yielded conflicting results in human studies in different laboratories. In 1986 Toma and Curtis reviewed recent reports in which some scientists reported no significant effects on iron and zinc absorption with 22 grams of wheat bran per day. Others reported significant decreases in iron and zinc balances at 35 grams per day. In general, most of these studies found that levels of wheat bran lower than 20 grams did not create negative balances of iron, zinc, and calcium in humans; especially if phytic acid is held constant and factors that promote iron absorption (vitamin C, fresh fruits, vegetables, and meats) were present in the diet.

Kelsay in 1981 reviewed many earlier reports of fiber and mineral binding in humans and arrived at the same general conclusions concerning low-fiber diets and iron balances. Brown bread at levels of 40–50% of total calories, whole brown rice, wheat or corn brans, and fresh fruits and vegetables at levels of 26 grams (almost 1 ounce) per day a month seemed to have no effects on iron balances. But whole-meal wheat breads, purified cellulose, hemicellulose, wheat brans, and fruit–vegetable neutral detergent fibers (that is, the hemicelluloses, pectins, gums, and mucilages) at similar levels, or at even half these quantities, increased fecal loss of zinc and led to a negative balance. Corn bran had no effect on copper balances, but wheat bran seemed to improve them, perhaps because wheat bran contains some copper. Cellulose, hemicelluloses, and pectins from fruits and vegetables produced negative copper balances at similar levels.

The apparent conflicting results between different laboratories emphasizes the fact that mineral binding to dietary fibers is quite complex. It does suggest, however, that persons who consume high-fiber diets should consider taking a mineral supplement or consume mineral-rich foods to avoid excessive losses. This precaution could be important in certain areas like Mexico or South America where the primary diet staple is corn (maize) served as corn tortillas with cooked beans, which contain phytic acid. Corn plus beans can make up to 66% or more of total calories in these diets. Animal proteins and other vegetables, whenever available, should be added to these diets periodically to help offset potential negative mineral balances.

PECTINS

Dietary fiber of fruits and vegetables consists of cellulose, hemicelluloses, and lignin (present also in cereal brans), but is primarily a good source of pectins. We've seen that the chemical composition of pectins is quite different from the other fibers. So, as might be expected, their binding properties are also different. I've mentioned that pectin has minimal effects on iron but can affect zinc and copper balances in humans. Kelsay and her co-workers in 1979 conducted 26-day studies on the effects of fruit and vegetable fibers on zinc, copper, and phosphorus balances in humans. They measured mineral intakes as well as fiber contents for both low-and high-fiber diets. Zinc and copper balances differed significantly on the two diets, but phosphorus levels were unaffected. For low-fiber diets, fruit and vegetable juices replaced the whole fruits and vegetables. On the low-fiber juice diets, zinc and copper remained in positive balance, but on the high-fiber diets, they showed slight negative balances. The negative balances were attributed to the higher lignin and cellulose contents from cell wall materials of whole fruit, not necessarily to the pectins, which are still slightly present in the juices. Fruit and vegetable fibers contain more lignin and cellulose but less hemicellulose than would be found in equivalent amounts of fiber from cereal brans. Brans also contain more magnesium, zinc, and copper than this particular diet containing fruits and vegetables. Thus it appears that the fruit and vegetable pectins are not as active as cereal celluloses, lignin, and hemicelluloses in binding to minerals. But pectins do not contain minerals like the cereal brans, so that diets high in fruits and vegetables are not as good a source of minerals as those with cereal grains.

Pectins have been found to bind to poisonous heavy metals, which could be a very desirable property. Researchers in Scotland (Rose and Quarterman, 1987) showed that alginic acid, cellulose, pectin, agar, and carrageenan all bound lead and cadmium to varying degrees. Alginic acid had the greatest binding to lead, and carrageenan bound the most cadmium. The loss of these toxic heavy metals via excretion of the fiber–metal complexes would be beneficial to the host.

PHYTIC ACID

Phytic acid's effects on fiber–mineral interactions deserves separate mention because of its presence in whole cereal grains and dried beans. The effects on Iranian males who ate brown bread were noted earlier, but phytic acid added to white bread also interferes with iron absorption. Zinc uptake from whole wheat or wheat bran products is also less than that from refined wheat products, suggesting phytic acid in the bran as the likely binding agent. The leavening process, however, appears to hydrolyze most of the phytic acid present in whole wheat products, thereby releasing bound minerals such as calcium, magnesium, and zinc for maximum uptake.

Phytic acid (also called phytin or phytate) is widely present in nature. It is found in all plant seeds, in a few root and tuber crops (carrots, parsnips, potatoes), but not in onions, turnips, green leafy vegetables such as spinach, cabbage, celery, or lettuce, or in fruit pulps. Highest concentrations are found in cereal brans and legumes, especially dried beans. The fiber from cereal brans and dried beans seems to bind to metals more than other forms of fiber, which tends to implicate phytic acid.

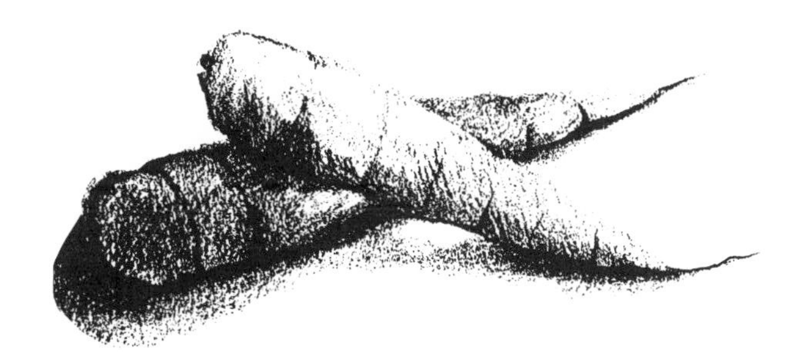

In our studies on the binding of minerals by rice hemicelluloses, we also examined the effects of phytic acid on binding of calcium and magnesium (Mod et al., 1982). Phytic acid had a profound effect on both minerals. As the concentration of phytic acid was increased from 0 to 33% to 50% in the reaction mixture, the calcium bound to phytate (instead of to the hemicellulose) and went from 0 to 19% to 41%, respectively. In the magnesium tests, as phytate concentration went from 0 to 33% to 50% the amount of magnesium bound to phytate went from 0 to 13% to 23%, respectively. Similar results of mineral binding in phytic acid–fiber tests have been reported by many laboratories. In all cases, addition of meat or mineral supplements to the diet readily restored the positive balance.

Zinc appears to be the mineral most easily affected by high fiber or phytic acid in the diets. This effect has been demonstrated repeatedly in numerous laboratories. These studies have emphasized that yeast fermentation of bread dough during the leavening process significantly increased the physiological availability of zinc from the bread by providing enzymes that can release this mineral from phytic acid.

♦

With so many of today's processed foods containing purified cellulose or wood pulp in high-fiber breads; as a binder in sausages; and as a thickening agent for soups, sauces, frozen desserts, and puddings, it is especially important that quality protein be included in a balanced diet to ensure that the mineral intake is optimal. Adding 10–20 grams of fiber per day to a balanced diet should be considered a safe level to avoid negative mineral balance for adults in good health.

The Modern Genie of the Can
The days of magic are gone.
The wizardry of yesterday is the reality of today.
Aladdin's lamp has long been lost—but its genie still lives in the spirit of modern science and industry.
In this sense, the Del Monte can is a magic container that annihilates distance and merges all seasons into one long fruitful summer.
Like the genie of the fable, it is ever at your command—ready to serve you the delicious products of the world's finest orchards and gardens—at any moment—on all occasions.
Just say Del Monte to your grocer and you may serve what you will—luscious, ripe pineapple from far-off Hawaii—golden, full-flavored peaches and succulent pears—ruddy, tree-ripened apricots, juicy plums and sweetest cherries from California's fairest valleys—famous Santa Clara prunes—raisins from Fresno—Oregon apples and berries from Utah—tenderest asparagus from the Sacramento's fertile delta—red-ripe tomatoes, delicious young peas, beans, spinach, pumpkin and squash—or any of the many other delicious varieties in the broad Del Monte line.
All are grown where they attain their finest flavor. In each is that natural goodness preserved intact.
This is the magic of the Del Monte can.
All its wizardry is summed up in the far reaching influence of the Del Monte ideal—in the thoroughness and care, the long years of experience, the scientific equipment and the unremitting attention given by experts to every stage in the growth, harvest and preparation of Del Monte products from the planting of the seed until the perfectly flavored fruit or vegetable is ready to be served on your table.
CALIFORNIA PACKING CORPORATION
SAN FRANCISCO, CALIFORNIA
Del Monte
BRAND
QUALITY
CANNED FRUITS
AND VEGETABLES

8

Is Fiber an Essential Nutrient?

The six essential nutrients are protein, carbohydrate, fat, vitamins, minerals, and water. Of the six, only protein, carbohydrate and fat contribute calories to the diet. Protein and carbohydrate contribute 4 calories per gram (1 ounce equals about 28 grams), and fat contributes 9 calories per gram. In planning balanced meals, people must also take into account the nutrient density of a food item. The *nutrient density* relates the amount of nutritional value in a food to its total calories. Thus, preparing a balanced meal high in protein and/or carbohydrates and low in fat will increase the nutrient density of the meal.

HOW WE EAT

A few decades ago, nutritionists and food processors could not sell nutrition on its own merits. Foods had to look, smell, and taste good to capture the consumer's interest. When the Del Monte Corporation published the first advertising message for a food product in the *Saturday Evening Post* on April 21, 1917, only the quality of the product was emphasized. Later the labels on canned fruits and vegetables included information on nutritional value that contributed toward a tasty, convenient, well-balanced meal. As the scientific and business communities increased their emphasis on nutritional quality, consumers increased their interest in better health and physical well-being through improved nutrition and its role in better health. Today, nutritional labeling includes the amounts of the major nutrients, what percentages they provide of the government's recommended daily allowance (RDA) of various nutrients, plus additional information on items such as sodium, sugar, and dietary fiber. Some even include recommendations for physical exercise along with good nutrition practices.

Today many Americans eat meals away from home, especially in fast food restaurants. The story is told of a school teacher who was trying to instill an understanding of good eating habits in her class of 9-year-olds, and asked them what were the four basic food groups. One child raised a hand to give the quick answer, "Burger King, McDonald's, Kentucky Fried Chicken, and Pizza Hut". Those meals tend to be high in fat and sugar, and sometimes include an alcoholic beverage. Alcohol supplies no nutrients, but it does add calories. Today's consumers eat fewer fresh fruits, vegetables, whole-grain breads, and cereals but eat more cream soups and sauces, deep-fried chicken and fish, French fried potatoes, soft drinks, and fatty meat dishes in which the fat has not

been removed. However, interest in fiber is becoming more evident as more restaurants, including the fast food services, add salad bars. Though dietary fiber is not considered an essential nutrient, it does play an important role in maintenance of good health. More attention is being given to the amount and type of fiber included in the diet and its nutritional implications for metabolism and activity in the GI tract. Unlike the essential nutrients, fiber's benefits are more indirect than direct.

PROTEINS, STARCHES, AND FATS

The protein we eat, whether derived from animals (beef, pork, poultry, fish, milk, and eggs) or plants (whole-grain cereals, beans, oilseeds, nuts, and other vegetables), is broken down by the body's enzymes to produce amino acids, which, in turn, are absorbed and used to rebuild the body's muscle tissue, ligaments, connective tissue, major organs, blood, hair, fingernails, hundreds of new enzymes, and hormones. Without dietary protein, the body will break down its own muscle tissues to obtain the amino acids it needs; this breakdown creates malnutrition that, unless reversed, leads to deterioration and eventual death. Protein malnutrition is called kwashiorkor. On recent television documentaries and in photographs, this disease is evident in small lethargic children with swollen bellies, scaly skin, and stringy hair.

Starch (carbohydrate) can be derived from various plant sources. Cereal grains such as wheat, rice, barley, or corn (maize) and potatoes are the primary sources of carbohydrate in most of the world, but in some places like Africa and parts of Asia, starchy crops such as yams, cassava, tapioca, and sweet potatoes are the principal sources. Carbohydrates are the energy-producing foods. They are broken down by the body's enzymes into simple sugars that undergo a series of biochemical reactions to produce the fuel the body needs to function properly. If not enough carbohydrate is consumed, the body will use fat from the diet or from the body's own fuel depots of stored fat to get enough energy. When this fat is used up, unless new sources are taken in, the body will start to use up its own muscle tissues as a last resort. The lack of both sufficient protein and calories in the diet leads to a condition known as marasmus. Children or adults suffering from this form of malnutrition are badly emaciated and weak and generally do not live long.

Vegetables and fruits provide many nutrients. They are good sources of vitamins; some are good sources of minerals, protein, simple sugars, and maybe starch; almost all are excellent sources of dietary fiber. Like the fruits and vegetables, dietary fiber comes in all kinds and quantities. Whole-grain cereals are noted for their cellulose, hemicelluloses, and lignin. Fruits and vegetables are good sources of pectins, mucilages, and gums, although small amounts of each type of fiber may be present in all plant foods. Oilseeds, legumes, nuts, and leafy vegetables are other plant foods that supply dietary fiber.

Fat is the maligned member of the essential nutrient group. Health officials are trying to convince Americans to cut down their fat consumption, from 30–40% to 20–30% of their daily calorie intake, because of the adverse effects of high fat intake on heart disease, atherosclerosis, and various circulatory disorders. Emphasis, however, is on reduction of the saturated fats (such as animal fats, palm *kernel* oil, cocoa butter, and lard), and cholesterol.

Nutritionists have known that some fat, especially that containing mono- and polyunsaturated fatty acids, is essential for proper growth and development. Today we are finding out that oils containing large amounts of oleic acid, a monounsaturated fatty acid present in olive oil and peanut oil, are reported to have beneficial effects on heart disease. Epidemiological studies of populations in England, Finland, and Italy showed a much higher incidence of heart disease in England and Finland than in Italy, where people are heavy users of olive oil. People in England and Finland consume larger amounts of saturated fats from butter and fatty red meats.

We are also reading about the beneficial effects of fish oil containing omega-3 unsaturated fatty acids on heart disease and vascular disorders. These conclusions are based upon comparisons of Eskimos living under traditional conditions and Eskimos who moved to the West and adopted Western eating habits. The omega-3 fatty acids are essential for the synthesis of certain prostaglandins in the body, and for the prevention of certain diseases. This work emphasizes the fact that some fat or fatty acids are necessary for proper metabolism and growth in humans. It is the excess that has given fat its bad name.

VITAMINS, MINERALS, AND WATER

The noncaloric nutrients—vitamins, minerals, and water—present a somewhat different picture. Because of the many improved and fortified food products available today, vitamin and mineral deficiencies are virtually unknown; medical students must rely on photographs and case histories in order to study symptoms of vitamin deficiencies. Most food labels now list the contents and the RDA for the vitamins in processed foods.

Very briefly, vitamins are required for the prevention of night blindness and vision disorders (vitamin A); beriberi, pellagra, anemia, and improper metabolism of carbohydrate and protein (B vitamins);

scurvy (vitamin C); and rickets (vitamin D); and for the proper formation of prothrombin, the precursor of the blood-clotting agent thrombin, and the prevention of hemophilia (vitamin K). The exact function of vitamin E in humans, once called the anti-sterility vitamin on the basis of animal studies, is still unknown. Recent research has shown it has some physiological functions in specific enzyme systems and in the relief of muscular dystrophy symptoms, but many scientists feel that vitamin E's main function is to act as an antioxidant, that is, a compound that prevents oxidation in polyunsaturated fatty acids. Such oxidation gives rise to the development of rancidity and off-flavors in the fats and/or oils. Vitamin E is sometimes referred to as a "natural" antioxidant because it is present in all seeds that contain oil as a form of storage material (e.g., starch and protein). It is generally accepted that all cells require vitamin E.

Minerals are similar to vitamins in that they are needed in small quantities for specific purposes: calcium for proper bone and teeth formation, iron for hemoglobin synthesis and anemia prevention, copper for synthesis of hemoglobin and certain enzymes, phosphorus for bone and teeth formation. In addition, numerous enzymes in the body contain metals as part of their structure—iron, copper, zinc, selenium, magnesium, or manganese.

The blood and body tissues are mostly water. Water is the nutrient most taken for granted because most meals contain a beverage, and meals with soups, sauces, or gravy also contain some water. However, without sufficient water, food would not be metabolized properly, and feces would be hard and give rise to constipation and related disorders.

THE NEED FOR FIBER

Where does dietary fiber fit into the scheme? Even though elimination of fiber from a diet will not directly cause visible deficiency symptoms in humans, the indirect effects caused by a low-fiber diet suggest that fiber is definitely beneficial. But because there is still no definitive evidence that dietary fiber is an essential, irreplaceable component of the diet, it is not considered a "nutrient" in the usual sense. We still have no RDA by the Food and Drug Administration, but the National Cancer

Institute recommends 20–35 grams (about 1 ounce) per day. Most Americans eat less than half of this amount.

There is still no consensus for a single name for fiber. Dr. Gene Spiller, a long-time researcher on dietary fiber, lists several names for fiber in his 1986 book, *Handbook of Dietary Fiber in Human Nutrition:* dietary fiber, dietary fiber complex, plantix, plantix complex, purified plant fiber, unavailable carbohydrate, edible fiber, neutral detergent residue, and crude fiber.

Diseases resulting from deficiencies of essential nutrients can be cured only by taking the nutrients. In kwashiorkor and marasmus, reduction of the visible symptoms and return to normal weight may take several weeks after return of the missing nutrients to the diet. In vitamin deficiences, administering a missing vitamin can produce dramatic recovery in a few days or less, as many graduate students taking courses in nutrition have discovered during animal feeding tests. One major reason for not considering fiber an essential nutrient is that disorders like constipation, hard stools, and bowel irregularity can be corrected by taking nonfiber pharmaceutical preparations such as phenolphthalein-containing laxatives and mineral oils.

Today some pharmaceutical manufacturers are urging the use of hydrocolloid bulking agents ("unnatural" fiber) like carboxymethylcellulose and methylcellulose, which are not as harsh as chemical laxatives. They act in the same manner as gums and mucilages, by absorbing large quantities of water; they improve bowel motility and produce soft, bulky stools, which relieve constipation (as was discussed in Chapter 5). These products are usually recommended for persons who, for some reason, do not eat whole-grain cereal products, fruits, vegetables, and natural fiber-containing foods in sufficient quantities, or for those with severe cases of diverticular disease who must always keep the contents of the digestive tract soft to avoid further complications. The role of dietary fiber here is to prevent excessive pressure on the intestinal walls and the diverticuli by absorbing large amounts of water, thereby producing softer stools. Nevertheless, these indirect symptoms of fiber deficiency can be relieved by pharmaceutical preparations and chemical laxatives or substances other than sources of food fiber, so that fiber is not considered essential in the absolute sense.

Despite the lack of universal acceptance of food fiber as an essential nutrient, Dr. Spiller believes that the momentum is with those scientists who have found valid uses for high-fiber foods—treating diseases such as type II diabetes, preventing colon and rectal cancer and obesity, and controlling fat and food intake. The ultimate study on long-range effects of dietary fiber has not been undertaken yet. According to Dr. Spiller, such studies are complex because of the nature of the fibers. Unlike vitamins that are easily isolated or synthesized, dietary fibers are large, high-molecular-weight polymers that may change when separated from cell walls or when processed. Also, their effects may be altered in the GI tract.

Finally, unlike the essential nutrients that exhibit positive effects when administered to those suffering deficiency symptoms, fiber may seem to cause problems: intestinal discomfort, cramps, or excessive gas upon ingestion by persons who have been on a low-fiber regimen and are seeking to rapidly increase the kind and quantity of fiber in their diet. (Persons with extremely sensitive digestive systems, diverticulitis, or unfavorable conditions in the colon should consult their physicians before making any radical changes in their diets and then very gradually increase the amount and kind of fiber to allow the microorganisms in the intestinal tract to adjust to the new foods.)

It is also generally accepted that, for each decade after age 20, humans require from 2% to 8% fewer calories to maintain proper or normal health. The need for essential nutrients changes very little, if at all, however, so that increasing the intake of fiber can play a dual role that some people feel is essential for the elderly: preventing constipation by keeping bowel motility normal and altering nutrient density of the food ingested by decreasing the caloric intake.

SO, IS DIETARY FIBER AN ESSENTIAL NUTRIENT?

A nutrient, the dictionary tells us, is "a food, anything that nourishes; anything that promotes growth or development and repair of tissues in the body." Let's consider the things we eat. *Proteins* are broken down by proteases into amino acids that are used to build new body proteins, muscle tissue, enzymes, and hormones. *Carbohydrates* are broken down by amylases into sugars that are used primarily to provide the energy that powers all the body's physiological processes, enzyme reactions, and tissue breakdown and repair. Excess sugars are recombined to form glycogen (carbohydrate particles or granules) that is stored in the liver and the muscles (especially in athletes) for future quick energy needs. *Fat* is digested by lipase into fatty acids that are absorbed for resynthesis into the body's adipose tissue (fat depots) that cushion many of the major organs such as the heart or small and large intestines. In excess, it contributes to oversized waistlines, where it can serve as a source of energy when the weight loss programs are started. *Minerals* become part of bone, teeth, and essential enzymes for the body's biochemical processes. *Vitamins*, too, are critical parts of some of the body's enzymes and are necessary for many things, including proper eyesight, blood coagulation, bone formation, and anemia prevention. *Water*, possibly the most important nutrient, can make up almost three-fourths of the weight of the body's soft tissues. Dietary fiber, on the other hand, is not broken down by the body's enzymes (except for the hemicelluloses) to be reformed into body tissues or enzymes, hence, by strict definition, is *not* a nutrient.

Is fiber essential? Without repeating all that has been said in the previous chapters, I think the answer is an unequivocal YES! Fiber does not become part of the body's new tissues or processes, but without it, there is certainly enough evidence now that the body's growth and biochemical processes do not function at their optimum rate. The bloated, lethargic feeling of constipation is not natural, and the benefits of dietary fiber, Nature's own brand of natural laxative, in alleviating constipation are unquestioned. Without constipation, colon pressure is relieved, and the resultant disorders of hemmorhoids, diverticular disease, hiatal hernia, and varicose veins are preventable. Although colon cancer cannot be cured by increasing the intake of dietary fiber, there is now ample evidence that there is a lesser risk of its developing in people and populations who consume unrefined cereals and grains, fruits, and vegetables—all good sources of natural food fiber or roughage. We now have evidence that the growth and development of polyps (precancerous lesions) in the colon is significantly reduced by consumption of a high-fiber diet that is high in cereal bran. Obesity can be controlled by reducing the caloric intake, and dietary fiber is a logical means of reducing calories without lowering food volume or nutrient density. A comparable statement can be made pertaining to sparing the insulin needs of diabetics.

This evidence is why major health agencies like the National Cancer Institute now recommend that Americans reduce their consumption of fat, especially saturated fat, by at least 10% and increase their intake of dietary fiber to at least 30 grams (about 1 ounce) per day, choosing from a variety of whole-grain cereals, fresh fruits, and vegetables. Many combinations of good tasting, high-fiber foods meet these criteria, as is discussed in Chapter 11. For those who feel that everything is going well with their present high-fat, low-fiber diet that tastes so good that they feel no need to change it, just remember Murphy's Law No. 6: If everything seems to be going well as is, it means you may have overlooked something.

That something is *fiber*.

FIBER FLAKES
100%
Daily
Nutrition
NEW!
Super Grain
100%
BRAN

9

Breakfast Cereals

The "Battle of the Brans"

Not too many years ago—before most members of the household were rushing off to work or to school—breakfast was a leisurely meal, usually hot: eggs; bacon, ham, or sausage; grits or potatoes; and toast; prepared by the person who was not rushing off somewhere. Today, as life styles and eating habits have changed, food manufacturers produce convenient, attractive products that provide ample nutrition and require little or no preparation time. A 1987 Department of Health and Human Services survey reported that more that half of all adults eat breakfast every day, and 19% more try to eat breakfast sometimes. This trend has promoted a boom in the production of ready-to-eat breakfast products in the past 20 years, although the cereal industry dates back to the early 1900s. Today the breakfast cereals industry is valued at over $4.8 billion on the basis of sales of 2.2 billion pounds at an average price per pound of $2.18. The major cereal grains used in these products are corn, wheat, rice, oats, and barley. Although $2.18 per pound may seem high for crops that were formerly grown for animal feeds, the cost per serving averages less that 20 cents, which makes it an economical meal in addition to being quick, convenient and nutritious.

Most of the ready-to-eat cereals are fortified so that they can provide a good part of the day's vitamin and mineral requirements. Ready-to-eat cereals are available as flakes, puffed whole grains, shredded whole grains and brans, and granola-type products. They are not considered major sources of protein, but today's cereals are gaining in popularity as a primary source of dietary fiber, and not just in the United States but also in Europe. Breakfast cereals are growing in per-capita consumption more in the United Kingdom than in the United States. In the West German and French markets, they appear to be heading for 10–20% annual growth rates, as Caldwell reported in 1988.

With today's health-conscious consumers looking for "lite" in beer, salt, soft drinks, cookies, candies, canned fruits, puddings, frozen desserts, and confections, some cereals may fill another need. They are generally low in sugar, although some have aspartame or other non-nutritive sweeteners instead of sugar. And the high-fiber varieties have lower caloric values per serving than low-fiber cereals. It is understandable why the public's acceptance of ready-to-eat cereals is growing, and manufacturers continue to introduce new varieties every year to fill these needs.

COMPARING CEREALS

Breakfast cereal manufacturers are going all out to convince the public that their products are the answer to reduce the risk of dreaded diseases. It's a real "Battle of the Brans" in the supermarkets. The result is confusion over which cereal is really the best. Is one really better than all

"It's a new idea in high-fiber breakfast foods. When the cereal is gone, you eat the box!"

Reprinted with special permission of Cowles Syndicate, Inc.

others for all needs? Ask 10 people who eat bran cereals and you will likely get 10 different answers. The competition between manufacturers for this growing market is fierce. Until 1984, Kellogg's All-Bran, with 9 grams of fiber per serving, was the undisputed leader of the high-fiber bran cereals. After General Mills' Fiber One entered the market with claims of 11 grams of fiber per serving, All-Bran boxes sprouted new labels stating "With a full 10 grams of fiber in every serving". Fiber One followed with a bar graph on the back of the box to show 12 grams of fiber, but by late 1987, Kellogg's was producing All-Bran with Extra Fiber, containing 14 grams of fiber per serving. And the race goes on!

To be considered a high-fiber cereal, a cereal generally has a fiber content of 3 grams or more per serving. Most breakfast cereals are not up there in the 10–14-gram range, so their advertising emphasizes other attributes along with the health claims and quotes of the National Cancer Institute. These claims ("Eat more fiber . . . fiber's links to healthier digestion, reduction of the risk of cancer, diabetes, obesity, hypertension . . .") are factual but can be confusing. Statements such as "No bran flake has more fiber than Corn Bran" are true: wheat bran flakes have 4 grams per serving and Corn Bran has 5. However, no comparison is made to the shredded high-fiber cereals with 10–14 grams per servings. Another claim states "Post Natural Raisin Bran has the highest fiber bran flakes," which is also true, because all of the bran flakes have 3–4 grams per serving. On Nabisco Shredded Wheat 'N Bran boxes you read: "The only bran cereal with no added sugar and no added salt". A serving has the usual 4 grams of fiber per serving but none of the added sugar and salt usually found in others, a plus for those seeking to control their sugar and salt intakes.

But the purposes of advertising claims on the boxes are to stress the health claims for fiber in their product and to encourage further sales. It is not possible, therefore, to say that any one brand is better than all others for all people. It depends on the health of the individual, the remainder of the diet being consumed, and the particular benefits desired from the fiber. Table 6 compares some of the nutritional data on 16 high-fiber breakfast cereals available in supermarkets today. Vitamin contents are not shown because most cereals provide a large portion of the recommended daily allowance. You should know that those containing dried fruit (usually raisins) and nuts usually list a serving size larger

Table 6: Nutrition Information on One Serving of Some High-Fiber Ready-To-Eat Breakfast Cereals

Product (Serving Size[a])	*Calories per Serving*	*D.F.[b] (grams)*	*Total Carb.[c] (grams)*	*Sugar (grams)*	*Sodium (milligrams)*
General Mills					
Bran Muffin Crisp (1.4 oz.)	130	4	30	11	250
Clusters (1 oz.)	100	3	19	7	160
Fiber One (1 oz.)	60	12	21	2	230
Kellogg's					
All-Bran (1 oz.)	70	10	22	5	260
All-Bran, Extra Fiber (1 oz.)	50	14	22	0	140
Cracklin' Oat Bran (1 oz.)	110	5	20	7	140
Mueslix Bran Cereal (1.45 oz.)	130	5	31	11	100
Raisin Bran (1.4 oz.)	120	4	30	10	220
Raisin Squares (1 oz.)	90	3	22	5	0

that the usual 1 ounce (28 grams). Also, added fruit and nuts increase the caloric value of the serving (except for Post Raisin Bran) by 30 or more calories. The total carbohydrate contents of all of these cereals are fairly close (20–30 grams), but the sugar contents can vary from zero (All-Bran-Extra Fiber and Shredded Wheat 'N Bran) to 11 grams (Bran Muffin Crisp and Mueslix Bran Cereal).

It is interesting to compare the two raisin bran cereals, Post's and Kellogg's, which are supposedly the same type of product. The dietary fiber contents are the same (4 grams), but the sugar, sodium, calories, and carbohydrate contents are higher in the Kellogg's product.

Like sugar, the sodium (salt) contents also vary, from zero in Raisin Squares and Shredded Wheat 'N Bran to 300 milligrams in Corn Bran and Bran Chex. For those suffering from risk of hypertension and/or heart disease, this information is useful.

Table 6: Nutrition Information on One Serving of Some High-Fiber Ready-To-Eat Breakfast Cereals *(Continued)*

Product (Serving Size[a])	*Calories per Serving*	*D.F.[b] (grams)*	*Total Carb.[c] (grams)*	*Sugar (grams)*	*Sodium (milligrams)*
Nabisco					
100% Bran (1 oz.)	70	10	21	6	190
Fruit Wheats (1 oz.)	100	3	23	5	65
Shredded Wheat 'N Bran (1 oz.)	110	4	23	0	0
Post (General Foods)					
Natural Bran Flakes (1 oz.)	90	5	23	5	240
Natural Raisin Bran (1 oz.)	90	4	22	4	180
Quaker					
Corn Bran (1 oz.)	120	5	24	6	300
Ralston					
Bran Chex (1 oz.)	90	5	24	5	300

[a]*The serving size of 1 ounce is 28 grams or one-half cup. Larger sizes of 1.4–1.45 ounces contain 1 ounce of cereal plus 0.4–0.45 ounce of raisin or other fruits.*
[b]*D.F. is dietary fiber.*
[c]*Total Carb. (carbohydrate) is combined starch, related compounds, sugar, and fiber.*
Source: All data were taken from ingredient labels on the boxes.

At this point, it is interesting to compare values in Table 6 with values determined by Baker and Holden in 1981. They published a report on the neutral detergent fiber (NDF) contents of 81 cereal products, which included eight of the brands listed in Table 6. They analyzed 3–12 samples of each brand (420 total samples) and reported the data as percentage of neutral detergent fiber as eaten. The two raisin brans list 4 grams dietary fiber per serving on the labels, or 14.29%, but Baker and Holden found NDF values of 8.4% for Kellogg's Raisin Bran and 10.5% for Post Raisin Bran. They found similar differences between Quaker Corn Bran and Ralston Bran Chex. Both list 5 grams of dietary fiber on the labels, or 17.86%, but Baker and Holden reported 18.1% NDF for the Corn Bran and 15.4% for the Bran Chex. These findings emphasize a point made in an earlier chapter: that the methods used to analyze fiber samples can yield different results, especially when analyzed by different people. The method used to obtain dietary fiber values on the box ingredient labels is not identified, but is probably a standard total dietary fiber method. The NDF method yields values that may differ slightly from the TDF values, but the differences in the NDF values cited here are not easily correlated with the dietary fiber values on the boxes. A look at the other ingredients in the products shows some differences, which may possibly explain the differences in NDF values found by that method.

Two cereal products released into national distribution are based upon whole oats. Kellogg's Nutrific Oatmeal consists of raisins, rolled oats, whole-grain barley, wheat bran, etc., and has 6 grams of fiber per serving. Quaker Oat Squares consists of whole oat flour, whole wheat flour, etc., and contains 2 grams of fiber per serving. Both have very good flavor but obvious differences in their amounts of fiber.

BRAN AND CEREALS

If the major benefit sought is prevention of constipation, then the high-fiber wheat bran cereals are the product of choice. We should not forget, however, that people eat food—not fiber. The current interest in fiber and its health benefits has dramatically changed our eating habits since the 1960s. However, not everyone can tolerate the coarse, high-fiber products that contain 10–14 grams per serving of wheat bran as their primary ingredient, because the coarse particles pass through the intestines largely undigested. In people who have severe diverticulitis, these particles may be trapped in the "pockets" (diverticula) and cause considerable abdominal pain. For such persons, bran flakes or products containing more soluble forms of fiber (such as oatmeal or oat-based products) may be more desirable.

Wheat bran or whole wheat grains are used in most of the high-fiber breakfast cereals because the bran is readily available from the refining of wheat, it is relatively cheap, and its water-holding capacity is the highest of the available cereal brans. From the corn milling industry, corn bran has also become available for cereals. The water-holding capacity of corn bran is less than that of wheat bran, but (from personal experience) a large bowl of corn bran for breakfast can be as efficient as wheat bran in laxation.

Wheat bran consists of lignin, hemicelluloses, and cellulose, the insoluble forms of fiber that have excellent water-holding capacity. (see the sketch of a cereal grain). In general, wheat, rice, and barley grains have this same basic structure: a large starchy endosperm contained in cells consisting of hemicelluloses and cellulose in the cell walls surrounded by the bran (aleurone) layer, one to three cells thick. The cells contain protein bodies, oil droplets consisting of polyunsaturated fatty acids, vitamins, and minerals. The cell walls are composed of hemicelluloses, cellulose, and lignin, with small amounts of pectins. On the tip of the grain that connects to the plant is the embryo, or germ, which is like the bran in that it consists of protein, more oil, vitamins, minerals, fiber, and many more enzymes needed to digest the storage materials in the seed if it germinates to form a new plant. When you eat whole grains or whole-grain cereals, you're ingesting all of those nutrients. When the grains are milled to remove the bran and germ layers, these nutrients are lost, leaving only starch and a small amount of hemi-

"The box says it contains five times more roughage than the leading bran cereal."

Reprinted with special permission of Cowles Syndicate, Inc.

celluloses. Several decades ago most of the milled byproducts were fed to animals. Fortunately, today the milled bran and germ are recovered for incorporation into breakfast cereals and some granola-type products, or sold as separate products, such as wheat germ and oat bran.

The amounts of fiber differ between the cereal grains and can differ between varieties of the same cereal. Hashimoto and co-workers in 1987 measured 6.7% fiber in whole wheat, 7.1% in whole barley, and 3.1% in whole rice. The extracted bran layers were double those amounts; wheat bran had 13.3% fiber, barley bran had 16.4%, and rice bran had 6.8%. In our research on rice, we found the following different neutral detergent fiber contents between long- and medium-grain rices and between the sources of fiber:

Source	*Bran NDF(%)*	*Endosperm NDF(%)*
Starbonnet long grain grown in Louisiana	29.7	2.9
Starbonnet long grain grown in Arkansas	34.2	2.7
Calrose medium grain grown in California	44.7	2.4

It is necessary, therefore, when describing a cereal grain, to identify the variety and perhaps the growing area, because there are obvious differences in their fibers. With rice, the medium-grain varieties appear to have more NDF than the long-grain varieties, and also different bile acid binding properties (described in Chapter 3).

Rice bran has not been used as much as wheat bran by the cereals industry because rice oil is a good polyunsaturated oil that tends to break down and oxidize quickly after milling. There are several active enzymes in the rice bran and in the bacteria and fungi adhering to the grains. As soon as the bran is removed, these enzymes begin to hydrolyze the fatty acids from the oil. However, a stabilization process developed by scientists at the U.S. Department of Agriculture's Western Regional Research Center in California inactivates the enzymes, stops the formation of free fatty acids, and stabilizes the bran, thereby extending its shelf life. This stabilized rice bran should find new applications in foods such as cereal products. That's good news because rice products are hypoallergenic (contain no gluten as in wheat, which causes celiac disease), readily digestible, and the bran is a good source of complex carbohydrates and fiber.

Barley is a good source of dietary fiber, but most of the U.S. crop is used in brewing beer. It is not currently a major source of dietary fiber because of a lack of continuous availability in large quantities.

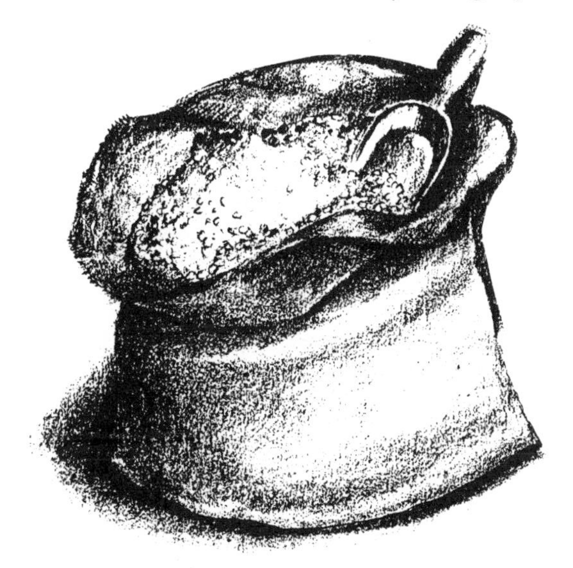

Oats, like most cereals, was once a major animal feed, primarily a high-grade feed for horses. As a result of the current interest in fiber and in research that has shown oats to have a total fiber content of about 9.0–9.6%, consisting of both soluble and insoluble fibers, interest in oats as human food has grown beyond the traditional cooked rolled oats for breakfast. The total amount of fiber in oats is lower than that in wheat, but the soluble fiber appears to have quite different properties that are more beneficial in lowering serum cholesterol. Oat bran is high in a gum called beta-glucan, a soluble fiber with a high affinity for the bile acids that act as precursors for cholesterol synthesis. This glucan is also responsible for the typical texture of cooked oatmeal and its viscous nature. For those seeking a means of lowering blood cholesterol levels, 2–3 ounces of oats per day as cooked oatmeal, cookies, or muffins, plus a lowered intake of saturated fat, can be quite beneficial. According to the American Heart Association, for every 1% drop in blood cholesterol levels, there is an estimated 2% drop in the risk of death from heart disease.

Seibert in 1987 cited a study of hypercholesterolemic men in which 94 grams (3.3 ounces) of oat bran per day was found to reduce serum cholesterol levels an average of 13%. LDL cholesterol, the bad type, was lowered by 14%, and the good HDL cholesterol was unaffected. New oat bran breakfast cereals started to appear on store shelves late in 1988. Oat bran, the newest "miracle (?)" fiber, is now the first ingredient listed in Kellogg's Cracklin Oat Bran (4 grams of fiber per serving), Common Sense Oat Bran (3 grams), and Ralston's Oat Bran Options (3 grams). Other new breakfast cereals listing oat bran as a second, third, or additional ingredient are Post Oat Flakes (2 grams), Kellogg's Mueslix (4 grams), Ralston's Bran Chex ("now with oat bran", 4 grams), and Cheerios (now listed as "an Excellent Source of Oat Bran", 2 grams). And you can bet that more are sure to come.

♦

It should be obvious, then, that the ready-to-eat cereals can play an important role in providing a good, fiber-rich breakfast *if*

1. The individual can tolerate the high-fiber wheat brans.

2. The intestinal walls are not irritated and diverticula are absent.
3. Other foods and amounts and types of fiber are eaten.

You need also to decide which effects of fiber you desire:

1. Prevention of constipation
2. Fiber's water-holding and bulk-forming capacities
3. Removal of bile acids and toxicants
4. Decreases in cholesterol and triglyceride levels in the blood
5. Weight loss or control of obesity
6. Diabetes control and insulin management
7. All of the above

If laxation or prevention of constipation is the major goal, the wheat bran cereals are best. If lower cholesterol levels are the goal, oat bran is the choice. If other forms of fiber are included in the daily diet to meet the major health agencies' recommendations for 20–30 grams of fiber per day, any products that provide 5 grams or more per serving can be beneficial and will help break the monotony of eating the same food every day. Despite a lack of research on human feeding studies with ready-to-eat breakfast cereals, they are considered good sources of fiber that make eating breakfast easier to do, even for weight-watching dieters.

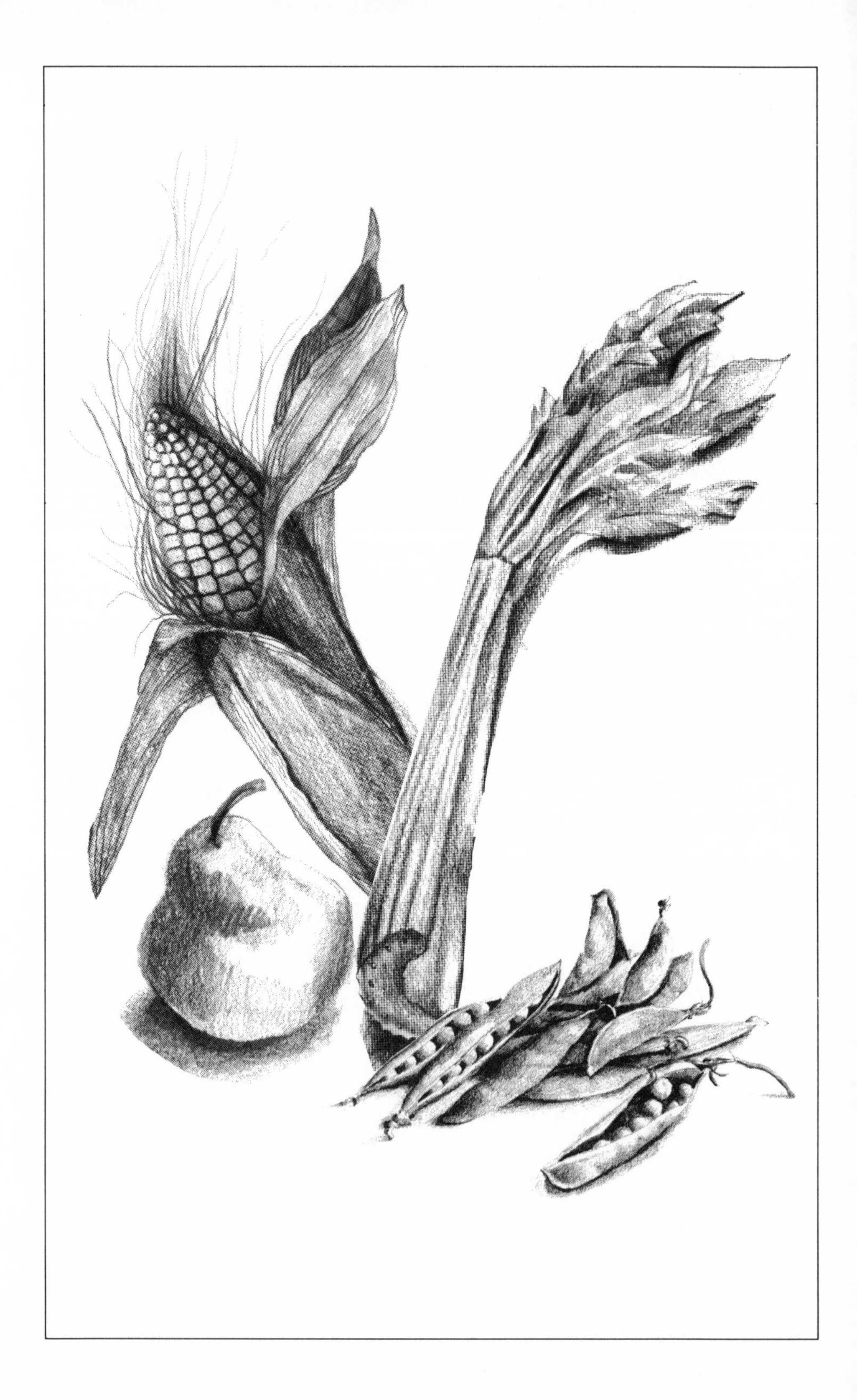

10

Fiber in Fruits and Vegetables

After all the advertising hoopla urging you to jump on the "bran wagon", it might appear that bran is the *only* fiber to eat and the breakfast cereal manufacturers have all that you need. However, recall what Grandma said half a century ago: "a starch, a protein, a green and yellow vegetable, and roughage." She did not emphasize only whole cereal grains, but included vegetables in a balanced diet. If we go even further back in history, we can read in the Bible of Daniel the Prophet, who may have been the first food fiber advocate to recommend vegetables over meat for improving health. In Daniel (1:8–16), we read

> But Daniel resolved that he would not defile himself with the king's rich food nor with the wine which he drank . . . Then Daniel said to the steward . . . "Test your servants for ten days, let pulse [beans] be given to us to eat and water to drink. Then look upon our faces and the faces of the children who eat the king's rich food, and according to what you see, deal with your servants." After ten days it was seen that they were better in appearance and fatter in flesh than all the children who ate of the king's meat. So the steward took away their rich foods and the wine they were to drink and gave them pulse.

It is evident, therefore, that centuries before our modern-day nutritionists (and Grandma) began to extoll the virtues of food fiber and vegetables, the ancients also realized that vegetables contained something that was good for health. With our current technology for analyzing foods and conducting feeding tests with animals and humans, we now know what the ingredients are, their chemical compositions, their reactions and interactions with other metabolites in the body, and the beneficial health effects that can be attributed to their inclusion in the diet. This is not to say that everyone should become a vegetarian but that meals should be balanced, not consisting of animal products alone to the exclusion of plant foods. We have also found that neither cereal brans alone nor beans alone have all of the desirable properties. No *one* source of fiber does all things for all people.

Despite all the interest in fiber and the evidence for its beneficial effects, not all of the health agencies making dietary recommendations between 1980 and 1985 recommended eating more fiber. Those that did were the American Cancer Society (1984), the National Cancer Institute (1984), the American Institute for Cancer Research (1982), and the U.S. Department of Agriculture–Department of Health and Human Services (*1985 Revised Dietary Guidelines*). The National Academy of Sciences Committee on Diet, Nutrition, and Cancer (1982), the National Academy of Sciences Food and Nutrition Board (1980), and the American Council on Science and Health (1985) made no recommendations (prior to 1986 when Pariza's report appeared).

The primary components of the dietary fiber in fruits and vegetables are the pectins, hemicelluloses, and gums. The chemistry and the properties of these fibers were described in Chapter 2 and will not be repeated here. Unlike the cereal brans that contain more cellulose and lignin, the fiber in fruits and vegetables can be modified by excessive cooking or processing (such as canning). In addition to losses of vitamin C and some B vitamins, some of the pectin can be degraded. Plat and co-workers in 1988 studied changes in the pectins of carrots before and after cooking. They measured an increase in the ratio of neutral sugars to uronic acids, from 0.11 to 0.27, an indication of some pectin degradation. Beans, an excellent source of fiber, are also affected by cooking. Monte and Maga in 1980 examined the effects of cooking on soluble and insoluble fibers from pinto beans. Cooked beans contained twice the amount of soluble fiber than raw beans, but cooking reduced the extractable hemicellulose-A by a third and completely depleted the hemicellulose-B. Extraction by their technique showed a 50% decrease in lignocellulose and crude cellulose after cooking.

Whereas carrots can be eaten raw, beans and peas cannot be eaten without cooking—heat is needed to inactivate the enzyme inhibitors that block the digestion of protein and starch in the digestive tract. Although cooking is necessary for some vegetables, overcooking should be avoided to retain integrity of the fiber and vitamins. When possible, you should buy fresh vegetables rather than canned products. Frozen vegetables have been blanched to inactivate enzymes before freezing, but are closer to the fresh state than canned products.

FIBER IN FRUITS

How may times have we heard that "an apple a day keeps the doctor away"? But no one really knew why the apple helped. The geriatric jokes about prune juice (or prunes) and its effects are also plentiful. Because of the current interest in fiber, however, analyses of dietary fiber in many other raw and processed fruits have been conducted in several laboratories. In addition, food labeling practices have prompted some processors to list fiber contents on their labels, especially dried fruits. To give you an idea of the variations between raw and processed fruits and between the various laboratories, dietary fiber contents of several common fruits are listed Table 7. As we see, the "apple-a-day" with the skin is a good source of soluble fiber. On a dry matter basis, the apple is equivalent to 5.2–7.9% fiber, depending upon the variety.

Walter and co-workers in 1985 examined the fiber content of apple pomace (the residue remaining after the commercial preparation of apple juice from fresh fruit). The crude fiber content (usually a lower value than total dietary fiber) of the pomace was 39%, a result suggesting that the dry matter in apples is a good source of pectin, hemi-

Table 7. Dietary Fiber Contents of Some Fruits

Fruit	*Fiber (%)*	*References*[a]
Apple, fresh with skin	2.0–2.2; 5.2–7.9	A, B, C, D
Apple sauce, canned	2.4	B
Apricot		
fresh	1.4–2.1; 7.1[b]	A, B, D
canned	1.5	B
dried, uncooked	12.01	B
Banana, ripe and raw	1.5–3.5; 4.9–7.6[b]	A, B, C, D
Blackberry		
fresh	6.7	A
canned	5.69	B
Blueberry	19.7[b]	D
Cherry, fresh	1.1–1.65; 2.1[b]	A, B, D
Coconut, fresh	13.64	B
Date, dried	8.75–8.88	A, B

celluloses, and cellulose. It also suggests that fruit juices will generally be quite a bit lower in fiber, except for trace amounts of soluble fiber.

Dried fruits (apricots, dates, figs, prunes, and raisins) have much higher fiber contents than fresh fruit. That's not just because they have lost a lot of water; there are also beneficial effects from certain enzyme-induced reactions during the drying process. Apricots (12.01% fiber) are also a good source of vitamin A, one of the vitamins reported to be beneficial in reducing the incidence of cancer. The "biggies" among dried fruit for fiber, however, are raisins, prunes, and figs.

Raisins are found as snack packs, in cookies, candies, breakfast cereals, raisin bread, raisin cakes, granola bars, and many other high-fiber food products. As Table 7 shows, raisins have almost five times more fiber than the grapes from which they are prepared. They contain both soluble and insoluble fiber, some of which is produced by enzyme-

Table 7. Dietary Fiber Contents of Some Fruits *(Continued)*

Fruit	*Fiber (%)*	*References*[a]
Fig		
dried, uncooked	16.99–18.5; 22.5[b]	A, B, D
canned	2.1	B
Grape, fresh	1.0–1.65; 1.6–1.7[b]	A, B, D
Orange, fresh	2.0–2.1; 3.4–3.7[b]	B, C, D
Peach		
fresh	1.4–2.3; 5.4–11.1[b]	A, B, D
canned	1.4	B
Pear		
fresh	2.3–2.5; 8.7–12.9[b]	A, B, D
canned	1.5	B
Prune, dried, uncooked	16.0–16.15	A, B
Raisin, dried, uncooked	6.81–7.1	A, B
Strawberry, fresh	2.16–2.29; 10.7–11.0[b]	A, B, D

Note: All values given are the range from the references cited.
[a]*References: A, author's files, 1982, B, author's files, 1987; C, Schweizer, 1987; and D, Spiller, 1986.*
[b]*Values are based on percent dry matter.*

catalyzed reactions during drying. They can be used whole or ground up into a paste to form a chewy center in certain breakfast cereals and bakery items. Raisins can also serve as a binder to hold other compounds in place in certain food products, and they have been reported to inhibit mold growth and thereby extend shelf life in various products due to their high content of propionic acid, a known mold-growth retardant.

Dried prunes, although not used as widely in food products as raisins, have been considered a rich source of fiber, and especially good for promoting bowel regularity, for decades. Even Grandma knew prunes were a good source of roughage, though she didn't know what kinds of fiber were present in them. The jokes about prune juice and the elderly are legion (Hear about the new geriatric cocktail? It contains four parts of prune juice and one part vodka, over ice.) But prune juice does not contain as much fiber as the whole fruit. The best way to consume prunes is as a snack in place of candies, cakes, and chips. With 16% fiber that consists of pectin, gums, hemicellulose, cellulose, and lignin, prunes are an ideal food to replace sweets and salted snacks. Like the other dried fruits, they are low in sodium and high in potassium and contain several vitamins.

If dried prunes are hard to chew (for elderly with dental problems), here's an easy way to soften them: fill a quart container about half full with the dried prunes, pour boiling water over the fruit to fill the container, cover tightly, and shake periodically over the next 2–3 days while storing the contents in a refrigerator. That will separate the prunes and allow the water to be absorbed. The prunes will then be soft enough to eat without much chewing, and the juice (containing soluble fiber) may be drunk. A single serving of prunes (four or five, depending upon the size) can provide 9 grams of fiber towards the day's requirements.

Soluble fibers, such as pectin, have been reported to have greater plasma cholesterol-reducing properties than the insoluble cereal fibers, except for oats, (according to a 1987 study by Somogyi). For people concerned about sulfite in fresh fruits and vegetables, only prunes, unbleached raisins, and dates can be successfully dried and stored without sulfite treatment or other chemical antioxidants (Somogyi, 1987).

Dried figs have been around for centuries, going back to the ancient Egyptians. But in recent years interest in figs has grown because of their high fiber content, almost 17%. This amount is more than most fruits, vegetables, or nuts. Figs can be used in most of the products in which raisins are used but they are not as appealing as prunes after rehydration and cooking in water. Canned figs are a minor nutritional item compared to the dried fruit. They are low in sodium and high in potassium and several B vitamins (although not as much vitamin A as apricots and prunes), and they are also a good source of calcium and iron. Like raisins, figs can be ground to a paste and added to a variety of baked goods to provide softness and moisture. Figs are natural humectants that can increase the shelf life of products in which they are incorporated. Like prunes, figs have also been reported to promote bowel regularity.

Fresh oranges are not considered a high source of fiber (about 2%), because most are squeezed to obtain the juice as a beverage. A smaller percentage of oranges are peeled and eaten whole; they retain the fiber as vesicles and cell walls. The orange peel is generally discarded, fed to animals, or extracted to recover the pectin. Braddock and Crandell, however, showed in 1981 that the orange albedo (the white portion of the peel) is an excellent source of fiber. Although it is only 3.3% of the whole orange, the albedo contains about 47% fiber components, mostly pectin, cellulose, and hemicelluloses. Some albedo is used in the preparation of crystallized fruit for incorporating into holiday fruit cakes but, in general, it is not used to a great extent in human foods. This is unfortunate; 50 years ago, my grandmother always saved the peel from oranges and grapefruit and prepared a tasty confection from it. (The recipe is given in Chapter 11.) Recovery of the albedo from citrus fruits could add a tasty new type of fiber to our diet.

FIBER IN VEGETABLES

Vegetables probably make up a greater proportion of our daily diets than fruits, so we should devote space to the pulses (legumes or beans) of Daniel the Prophet and to Grandma's green and yellow vegetables. The fiber contents of some of the more common vegetables are listed in Table 8. Table 8 illustrates how the variety of vegetables and the type and amount of processing all have an effect on the fiber content. In general, dry beans have more fiber than fresh (immature) beans, and cooking seems to increase the amount. This finding suggests that the New Orleans favorite dish of red beans (prepared from dried red kidney beans) and rice should be a high-fiber (and delicious) meal. However, if the dish is prepared with brown rice (2.46% fiber) rather than the usual white milled rice (0.74%), an additional amount of cereal bran fiber could be incorporated into the meal. Dried red kidney beans have been reported to have higher fiber than most of the other dried beans.

If you read the labels on canned foods, another American favorite, pork and beans, which is possibly prepared from dried white navy

Table 8. Dietary Fiber Contents of Some Vegetables

Vegetables	*Fiber (%)*	*References*[a]
Beans		
Broad, fresh, boiled	4.55	D
Green, canned, French style	3.28–3.4	A, B
Green, canned, cut	1.33–3.4	A, B
Lima, dried, mature	5.09	B
Lima, dried, canned-cooked	13.1–15.5	C
Beets, cooked	2.47	A
Broccoli, fresh or frozen		
Cooked	3.76–4.1	A, B
Raw	3.59	B
Brussels sprouts, fresh, cooked	2.9–2.92	A, B
Cabbage, green		
Fresh, cooked	2.47–4.0	A, B
Raw	4.0	B
Carrot		
Fresh, cooked	3.01–3.08	A, B
Raw	2.89	B
Celery, raw	1.74–1.83	B
Corn, sweet		
Canned	5.59	B
Cooked on cob	5.6	B
Eggplant, fresh		
Cooked in water	2.5	A, B
Fried in oil	0.95	B
Lettuce, fresh	1.45	A
Okra, seeds only, as crude fiber	27.2–27.3	E

Table 8. Dietary Fiber Contents of Some Vegetables *(Continued)*

Vegetables	*Fiber (%)*	*References*[a]
Peas		
Canned	3.2	D
Frozen, cooked	3.81	D
Potato, baked in skin	1.99–2.24	A, B
Rice, cooked		
Brown	2.46	A
White	0.74	A
Squash, fresh		
Winter	2.92	A
Zucchini	3.08	A
Spinach		
Canned	6.33	B
Fresh, cooked	3.66–6.29	B, D
Raw	3.34	D
Sweet potato, fresh, baked	2.38–2.8	A, B
Turnip greens, cooked	3.76	A

Note: All values given are the range from the references cited.
[a]References: A, anonymous, 1982; B, anonymous, 1987; C, Marlett and Chesters, 1985; D, Herranz et al., 1983; and E, Al-Wandawi, 1983.

beans, is being advertised as "Naturally High in Fiber, Low in Fat", with a dietary fiber content of 11 grams per serving. A serving is about 8 ounces (227 grams), half of the can. This amount of fiber could put pork and beans in the high-fiber class of foods with bran cereals. In general, all legumes, especially dried beans and peas, are low in fat, high in protein and fiber, and a rich source of minerals and vitamins.

The root vegetables (beets, carrots, potatoes, and sweet potatoes) and many of the green and yellow vegetables Grandma advocated generally contain about 2–3% fiber, but broccoli, corn (which has a bran coating), spinach, and turnip greens are higher. It is interesting to note, but more difficult to explain, that eggplant fried in oil loses more than half its fiber compared to eggplant that is cooked in water. Whether that loss is due to the higher heat of the oil or to another effect of oil is not known. In addition, the oil absorbed by the eggplant slices during frying is not recommended for the overweight or those with a tendency towards atherosclerosis or coronary disease. So it is much better to cook eggplants in water or steam or in a microwave oven rather than fry them.

Another vegetable, popular in the southern United States, Mexico, and South America, is okra, a crop that grows well in hot, humid climates. One of the major complaints about okra is the slimy–sticky exudate it produces during cutting and cooking. This sticky exudate, however, is a natural gum, one of the forms of dietary fiber present in the okra tissue and in its seeds (which are 27% fiber). This gum is desirable in the preparation of Creole okra gumbo, where it acts as a thickening agent. Okra can be prepared many different ways that remove the gum, if desired. In some places, okra is grown for the seeds, which contain about 20–21% protein and 14% oil. The protein is typical of most seed proteins, which are not as complete as animal proteins, but the oil has a good polyunsaturated fatty acid composition.

FIBER IN NUTS

Tree nuts and peanuts (which are not nuts, but legumes) are not considered good sources of fiber because of their usually high oil content. The oils vary in composition between nuts, but they are generally high in oleic and linoleic acids (the mono- and diunsaturated fatty acids), and low in saturated fatty acids. They contain no cholesterol, of course; cholesterol is an animal fat not found in seeds of plants. Tree nuts are not considered major sources of protein, but peanuts are crushed in some

countries (and to a lesser extend in the United States) to obtain the edible oil and a high-protein animal feed.

Table 9 lists the dietary fiber contents of four popular nuts. Peanuts with the red skins contain 3–4% more fiber than the other nuts. In an

Table 9. Dietary Fiber Contents of Several Nuts

Nut	*Fiber (%)*
Almonds, shelled or unshelled	5.30
Brazil nuts, shelled	4.12
Peanuts, roasted in shell, with skins	8.33
Walnuts, English, shelled	5.20

Source: Author's files, 1987.

earlier study (Ory and Flick, 1987), we analyzed peanut skins for neutral detergent fiber contents and found it was 43–45%. Skins represent only 2–4% of the weight of the shelled peanuts, so that eating the skins along with the roasted peanuts, as is done in peanut brittle, roasted–salted Spanish peanut snacks, "Cracker Jack"—the longtime snack favorite made of coated popcorn and peanuts—and roasted-in-shell "ballpark" peanuts, can provide another source of dietary fiber, even while snacking.

Dr. George Washington Carver, during his years of research on peanuts, developed more than 300 uses for peanuts, most of them in food. Because peanuts are legumes, it should be possible to prepare them in many of the same ways in which beans and peas are prepared.

◆

In summary, many fruits, vegetables, and nuts can provide ample amounts of dietary fiber. These forms of fiber are different from those in cereal bran, but just as important for our diet. These forms are generally more soluble than fibers from cereal brans, but their properties are beneficial in other ways. If breakfast belongs to the cereal brans, then lunch and dinner belongs to the fruits and vegetables. Together they can provide the balanced meals we require each day.

11

Can Fiber Be Good for You and Still Taste Good?

Eating has always been one of life's great pleasures. Some people eat to live; others live to eat. It is the latter group to whom this book is mainly directed.

Healthy diets need not be boring and tasteless. The old saying that if it tastes lousy, it must be good for you, is just that—an old saying, outdated. Healthy, balanced diets can be just as tasty as the wrong kinds of fatty, over-sweetened or over-salted foods that have generally been recognized as the most taste-appealing. We now have ample evidence that foods containing a lot of fat, salt, or sugar can increase the risk of certain types of cancer and other diseases, and can aggravate some types of digestive tract disorders.

Because of today's improved communications and technology and increased knowledge about proper nutrition, eating habits are changing for the better. We are more aware of the benefits of fresh fruits and vegetables, less fat and salt, and more exercise. But despite the improved knowledge, old habits die slowly. Results of a Louis Harris poll of 1250 adults in May 1988, conducted for *Prevention* magazine, showed that:

- 42% try hard to control intake of high cholesterol-containing foods
- 54% try hard to limit salt intake
- 54% try to limit fat, and
- 59% try to ingest enough vitamins and minerals.

These totals were down slightly (from 2–4%) from the 1987 poll. The only area of improvement in the poll was an increase in attempts to control stress in life (up from 64% to 68%).

STEPS TO GOOD HEALTH

The next step is to introduce proper nutrition. That is a bigger problem for the elderly because many live alone and don't have the energy or the inclination to prepare big meals. They are often on limited incomes, and they may suffer from chronic diseases that limit mobility or interfere with chewing or digestion. The physiological processes of aging reduce the senses of smell and taste; reduced senses increase the desire for foods that are extra sweet or salty, the "no-nos" of weight control and health maintenance. But proper nutrition need not be a chore or a boring task. Eating balanced meals can be easy, tasty, and can fit into most life styles from the very active to the more sedentary. (Just remember that the sedentary life style uses fewer calories, and the caloric intake should be adjusted downward to avoid increased weight and susceptibility to various disorders.)

"I don't mind eating healthier, but I refuse to eat prune lasagna!"

Reprinted with special permission of Cowles Syndicate, Inc.

There are several obstacles to acceptance of new dietary habits. Some are based on tradition ("We've been eating this way for generations"). As we age, we may find it hard to adjust to changing taste buds, physiological processes, and our decreased caloric output vs. caloric intake. Consumers are prejudiced against new foods and foods that are enriched with "chemicals", (although cereals are fortified with vitamins and minerals and are not perceived as having chemical additives).

All these factors can affect the acceptance and palatability of changes in diet. But proper education of the public with nutrition and diet articles in newspapers, magazines, television, radio, and in advertising of manufactured food products (but not "overkill" advertising), can make today's consumers more knowledgeable and more inclined to accept new products that taste good and are also nutritious. Some manufacturers underestimate the ability of consumers to understand advertising; they have overused the term "natural" on so many products that the public is beginning to doubt the reliability of such advertising.

We've about reached the overkill stage on uses of the "Big C" words (cancer, cholesterol, and coronary disease) on the labels of certain products. Some of this space could be used to identify which forms of fiber are present in the product, and which benefits have been related to or associated with these specific forms of fiber. This information would allow consumers to select fiber-containing foods for specific needs or goals. It would also help them to vary the types of fiber to get a balance that would help in several areas, rather than in only one. Americans are eating better, but there is always room for improvement.

Reading labels is an excellent means of selecting foods high in fiber, protein, vitamins, or minerals and low in fat, sugar, or sodium (salt), but there are rarely any nutrient labels on fresh fruits and vegetables, meats, fish, or poultry. This lack of information requires consumers to shop more wisely, but it is possible. For beginners, it would be advisable to plan menus for the week before going to the grocery store, and to bring a list of needs. Select several varieties of fresh fruits and vegetables (or frozen or canned, if that's more desirable for some reason). Choose whole wheat breads and crackers instead of milled wheat products. Select lean meats that have little or no fat (aged red meats have some "invisible" fat due to marbling, but this amount can be reduce by broiling or grilling). Select more poultry and fish. Eat fewer eggs and less cheese and sour cream. Use egg whites instead of whole eggs; use plain low-fat or nonfat yogurt instead of sour cream. Buy less bacon and processed meats like salami, bologna, and hot dogs. Eat brown rice instead of white milled rice.

Shopping wisely for good, nutritionally balanced meals is not more expensive if you buy lean cuts of meat rather than aged prime beef and processed meats. Poultry is also less expensive than beef. Fresh vegetables are generally less expensive than frozen or canned. Fresh fruits and dried raisins or prunes are generally less expensive per unit weight than candies, sugar topped cakes and cookies, and soft drinks. Remember also that alcohol contains empty calories and no fiber, under normal circumstances. (There was a man who drank so much wine that he ended up in the hospital suffering with malnutrition. The doctor told him he wasn't getting enough fiber in his diet and suggested some changes.

"But, Doc," he replied, "I don't like vegetables."

"Well," the doctor said, "You better start eating the corks.")

The best way to find out fiber contents is to read labels. Remember that the ingredients are listed in order of their concentration; the

item listed first is the major ingredient, and so on down the list. But don't stop eating dairy products completely. Calcium is a critical mineral, and dairy products are a major source of calcium. Select low-fat or skim milk, low-fat cheeses or cheese spreads, and low-fat or nonfat yogurt. Season food during cooking. Do not bring a salt container to the table. It is foolish to select low-sodium foods, then douse them with salt before eating.

ADDING BRAN TO FOODS

Sneak more fiber into low-fiber meals to reach the suggested level of 20–30 grams per day. For example, avoid the sweetened ready-to-eat cereals. Low-fiber cereals can be improved by adding a handful of raisins or a few prunes, figs, or apricots or some fresh fruit to the bowl. If you prefer hot cereals (oatmeal, wheat, barley, or rice meals), add a

handful of wheat bran, corn bran, or oat bran to the cooking pot. (For instant oatmeal, add the bran to the cereal before adding boiling water and mix well.) If you prefer pancakes or waffles, add bran to the batter and mix well before cooking. If a handful of bran seems like too much for one dish, add a spoonful or two to several items: sprinkle it on toast with jelly or peanut butter, in scrambled eggs or omelets, or on fruit salad as a sort of crouton. Using your imagination, you can surely find other acceptable ways to increase the fiber contents of a low-fiber breakfast.

If you prefer sandwiches, you can substitute whole wheat bread for white bread. Bran, raisins, or vegetables can be added to chicken or tuna salads and sprinkled, like croutons or bacon bits, on green salads, soups, or yogurt.

For evening meals many options are available. Bran can be sprinkled as toppings on baked potatoes, casseroles, vegetable lasagna, and scalloped or hash-brown potatoes. Bran can be added to meatballs, meat loaf, hamburgers, tacos, enchiladas, or burritos. For dessert, there are also many options. Add it to cookies, cakes, cake toppings and fillings, puddings, and fruit salad or use it as ice cream topping. Again, imagination can help to find other ways to add bran or raisins to foods low

"Eeny, Meeny, Miny, Mo,"
—say the children—
and they take what they get

Childhood's selection is left to chance! But it's hardly the way to make "grown-up" decisions. Especially in the choice of canned fruits. There are too many varying qualities, too many different brands, to trust to luck. You *must* know in advance the brand that guarantees satisfaction—then see that you get that brand. On canned fruits the name is DEL MONTE, every time—a sure guide to quality and flavor, no matter when or where you buy.

DEL MONTE

Reprinted from The Woman's Home Companion, December 1926

Reprinted with permission from Del Monte Foods USA, copyright 1926.

in fiber. The recipes in this chapter will illustrate some specific ways of preparing tasty, nutritious meals that are high in fiber from foods other than the cereal brans.

TEENAGERS AND FOOD CHOICES

Diet recommendations for children and adolescents vary from those for adults. Polls showed that more teenagers are eating breakfast today, a bowl of cereal rather than only coffee or tea or nothing at all. More teenagers are knowledgeable about nutritious eating habits and are concerned about developing severe health problems, such as heart disease or cancer, but they still observe poor eating habits. A majority still prefer potato or corn chips, cookies, candies, and other confections rather than a piece of fruit as a favorite snack. Many expressed concern about cholesterol and saturated fats, but a significant number still have eggs for breakfast at least three times per week, and a large percentage still favor hamburgers, cheeseburgers, pizza, and luncheon meat sandwiches for lunchtime meals. The traditional fat-filled favorites of a cheeseburger with bacon, a milkshake, and French fried potatoes still rank high on

the teenager fast food selection lists. But, fortunately, most fast food outlets now have salad bars that are gaining in popularity.

Even if the "traditional" items are desired, we can still limit fat and salt contents by selecting garnishes more carefully. For example, on pizza, choose mushrooms and sweet peppers without extra cheese instead of processed meats or sausage. If baked potatoes or rice are available, choose one of these instead of French fried potatoes—and don't add butter, sour cream, or grated cheese; use a dash of Tabasco sauce, black pepper, tomato catsup, or cottage cheese to flavor it. At the salad bar, eat lots of vegetables but delete the high-calorie dressings and bacon bits. Use less oil and more vinegar or lemon juice. Instead of croissants, choose buns or bread; croissants have almost five times more fat than two slices of whole wheat bread.

Another possible substitute protein for fast food meals is beans. Mexican restaurants offer refried beans, and one of the national fried chicken chains offers New Orleans style red beans and rice. Dried beans are low in fat and high in protein, fiber, vitamins, and minerals; they provide more protein than all other plant foods, and they have no cholesterol. Refried beans in tacos, burritos, or red beans and rice can provide ample protein in a meal and offer a change in flavors. For home cooking, dried beans are an economical source of protein (though not as "complete" as animal proteins), a good source of fiber, and they come in several varieties:

- black, white, red kidney, fava, lima, pinto, and soy beans,
- black-eyed, chick (garbanzo), green split, and whole yellow peas, and
- lentils.

All are good proteins that can be turned into tasty meals. For those concerned about the gas-forming sugars in dried beans, there is a custom used for generations in New Orleans that removes 75–80% of these sugars. Simply place the dried beans in a large bowl or pot and cover well (2–3 times their weight) with water. Add a small pinch of sodium bicarbonate to the water, mix well, cover the container, and let stand overnight in the refrigerator. Next day, pour off all of the water (which contains the extracted sugars), then cook according to the desired recipe.

Dietary Fiber Recipes

It is impossible to include all possible recipes for high-fiber meals, but the following are examples that allow for ease in substituting favorite and available foods for the vegetables or fruits listed. For breakfast, ready-to-eat cereals with fresh, dried, or canned fruits is the best recommendation. They are simple, require no preparation (except for cooked cereals like oatmeal), and provide the highest amounts of fiber. For lunch, soups, salads, or sandwiches on whole wheat breads can generally meet a person's needs. The evening meal allows more possibilities that may include items from the breakfast and luncheon menus but, again, each person should decide what is more appealing, tasty, and helps to reach the desired goals. If prune lasagna is your dish, then go for it.

SALADS

Salads are to high-fiber lunches and dinners as cereals are to breakfast. They are generally easy to prepare, they are low in fat (if not buried in salad dressings), they can be quite tasty and nutritious, and they provide a good portion of fiber for the body's needs. Salads can vary: simple tossed greens (lettuce, spinach, or watercress), lettuce and tomatoes, sliced beets on lettuce, or salads consisting of more combinations of vegetables. Some can be eaten with no dressing or simply a dash of lemon juice; others seem to require more complex or "creamy" dressings. However, creamy dressings need not be high in calories to be tasty. There are many salad dressings with less oil on the grocery shelves today, and some can be easily prepared at home. Here are some low-calorie salad dressings that can be made at home.

Yogurt (or Cottage Cheese) Creamy Dressing

½ cup plain yogurt (or low-fat cottage cheese)
2 teaspoons lemon (or lime) juice
2 teaspoons olive or peanut oil (optional)
2–3 drops Tabasco sauce (to taste)
½ teaspoon dried chives (or parsley) powder (if greenish color is desired)
½ teaspoon paprika powder (if reddish color is preferred)
½ teaspoon salt (optional)

If cottage cheese is used, add ½ cup buttermilk to make a thinner dressing. Mix all ingredients in a food blender until smooth. If too thick, add a teaspoon of water (or buttermilk) at a time until desired thickness is reached.

Dijon Mustard–Yogurt Dressing

½ cup plain yogurt
2 tablespoons Dijon mustard
2 tablespoons lemon (or lime) juice

Mix all ingredients in food blender or whisk them together until smooth. If too thick, add a teaspoon or more of water, and blend again.

Low-Calorie Vinaigrette Dressing

½ cup water
½ cup red wine vinegar
½ teaspoon salt (optional)
2–3 drops Tabasco (to taste)
2 tablespoons dried chives or parsley
2 tablespoons olive or peanut oil

Mix all ingredients except oil in a blender until smooth, then add oil slowly and blend for 30 seconds longer, until oil is well blended. Unused dressings should be stored in a glass or plastic container without a metal cover (because of the acidic nature of vinegar and lemon juice).

If desired, ½ teaspoon of garlic powder may be added to any of these dressings to increase flavor. Scientists are now finding out that garlic is more than a pungent flavor additive. Some garlic components have been implicated in the prevention of blood clots and the removal of certain carcinogens from the body. The ancient Egyptians used garlic 4000 years ago to treat heart disease, tumors, headaches, intestinal worms, dysentery, high blood pressure, and a host of other disorders. In 1858, Louis Pasteur found that garlic was also antibacterial.

In preparing all vegetable salads, fresh vegetables should be washed well in running water before using, especially if they will not be cooked before eating. Canned vegetables should be drained and added to salads without further cooking. Here are some multivegetable salads.

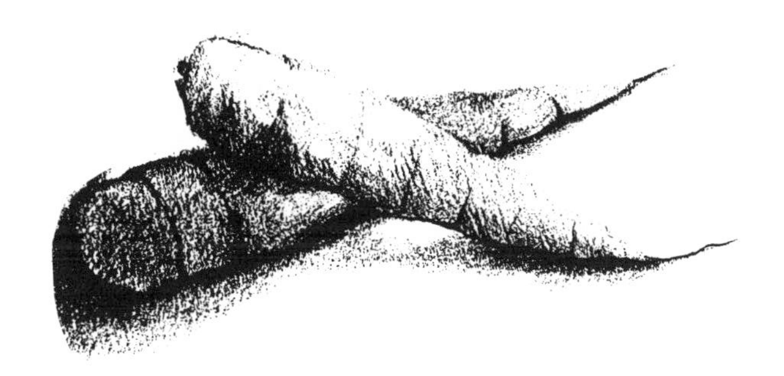

Deluxe Carrot Salad (my favorite)

3 large or 4–5 medium carrots, grated
2–3 pieces celery without leaves, finely diced
1 medium-to-large Delicious apple with peel, cored and diced
½ cup raisins
½ lemon, squeezed to obtain juice
2–3 tablespoons mayonnaise (or other salad dressing)

Place the vegetables and fruits in a large mixing bowl and pour lemon juice over them, mixing well to coat carrots and apples. (Lemon juice will block activity of the enzyme that causes browning.) Spoon the mayonnaise or other dressing over the salad and mix well again. Extra salad can be stored in an airtight container in the refrigerator for a day or two.

Waldorf Salad Without Pecans or Walnuts

2–3 Delicious apples with peel, cored and diced
½ cup raisins
2–3 pieces of celery without leaves, finely diced
2–3 tablespoons dry-roasted, unsalted peanuts, chopped (optional)
3 tablespoons lemon juice
2–3 tablespoons mayonnaise (or other salad dressing)

Place the apples, raisins, celery, and peanuts in a large bowl, add the lemon juice, and mix. Add dressing and mix again. Extra salad can be stored in an airtight container in the refrigerator for a day or so.

Deluxe Cole Slaw

½ head cabbage, medium size, shredded
1 green sweet pepper minus seeds, finely chopped
1–2 medium carrots, finely chopped
1 piece celery, finely chopped
1 green onion or shallot, finely chopped (optional)
3–4 tablespoons low-calorie creamy dressing
2 tablespoons lemon juice

Mix all ingredients well in a large mixing bowl. Excess can be stored in an airtight container in the refrigerator.

Fruit Salad

1–2 Delicious apples with peel, cored and diced
1 banana, peeled and sliced
1 orange, peeled and sectioned
1 8-oz. can pineapple chunks, drained (save juice)
3–4 tablespoons lemon juice
1 cup seedless grapes, sliced into halves or quarters
½ cup maraschino cherries, sliced into halves
½ cup strawberries, sliced (or whole blueberries, raspberries, or blackberries)
½ cup low-fat cottage cheese (or yogurt)
½ cup fruit juice (cook's preference: apple, orange, cranberry, plus the juice from the pineapple chunks)

Mix apples, bananas, orange, and lemon juice in large bowl to coat with the lemon juice. Add all other ingredients and mix well. (If you use yogurt instead of cottage cheese, use less juice.) Extra salad can be stored in an airtight container in the refrigerator.

(Another trick to keep fresh fruit salad from browning is to crush one or two vitamin C tablets, sprinkle over the salad, and mix well before storage.)

Three-Bean Salad

1 16-oz. can each of red kidney beans, black-eyed peas, and cut green beans (or cut yellow wax beans, lima beans, green peas—whichever are your favorites)
½ teaspoon garlic powder (optional)
½ teaspoon onion powder
1 medium purple onion, peeled and finely diced
1 green sweet pepper minus seeds, diced
1 piece celery, finely diced
½ lemon, squeezed to obtain juice

Mix the vegetables and herbs in a large bowl. Add lemon juice and oil and mix again. Serve cold on lettuce leaves.

Fresh Summer Vegetable Salad

2–3 carrots, scraped and sliced into ½-inch slices
2 ripe tomatoes, cut into ½-inch slices or pieces
2 small zucchini squash, sliced as for carrots
2 pieces celery, cut into ½-inch slices
2 small yellow squash, sliced as for carrots
1 broccoli stalk, cut into bite-size pieces
2–3 medium sweet peppers (a red, a green, and a yellow is nice) minus seeds, diced into ½-inch pieces
3–4 tablespoons olive or peanut oil
½ lemon, squeezed to obtain juice
½ teaspoon salt (optional)

Mix all ingredients well in a large bowl. Serve cold on lettuce or spinach leaves. Excess can be stored in an airtight container in the refrigerator.

SOUPS AND GUMBO

Soups are generally considered cold-weather meals that can contain whatever the cook has available or whatever the family likes: beef, poultry, pork, fish, barley, rice, pasta, and various vegetables. Louisiana's famous soup, gumbo, gets its name from the African word *ngombo*. It means okra, the vegetable used in preparing gumbo. African slaves brought okra to Louisiana, where its distinct flavor and other properties soon made it a popular vegetable for Southern cooking. The soluble fiber gums in okra made it ideal for thickening gumbo, which also included wild game, poultry, seafood, sausage, ham, or whatever else was available.

Soups are generally thinner than gumbos because they do not start with a roux as a base and do not contain a thickening agent. A roux is simply equal parts of oil and milled wheat flour, ½ to 1 cup of each depending upon the volume desired, that is cooked on a medium flame with constant stirring until it is dark brown in color. (In a microwave oven, the oil and flour can be heated uncovered at high setting for 6–7 minutes, stirred well, and returned to the microwave for 20–30 seconds longer.) To this base is added the "trinity": 1–2 cups chopped onion, 1–2 cups chopped celery, and ½–1 cup chopped sweet pepper, depending upon the final volume. These are added to the roux and cooked, stirring, until the onions are translucent. (In the microwave, add the trinity to the roux, mix, and cook at high setting for 3 minutes.) This is the general base to which cold water and other ingredients are added to prepare gumbo. For a soup, simply bring water or meat stock to a boil, then add the other ingredients.

Traditional Creole cooking has been based upon a good thick roux (mostly pork fat or bacon "drippings", definitely not the most heart-healthy). But a new Creole cuisine is emerging that uses less fat and more meat, fish, or vegetable stocks that are reduced to a pastelike consistency and used instead of the oil–flour roux. There is sufficient Maillard browning to give the desired color, to which the vegetable trinity can be added, and the particular recipe being prepared can then be followed.

Hearty Vegetable Soup

½–1 lbs. beef brisket trimmed of all fat and cut into small pieces (optional)
½—2 quarts meat or vegetable stock, or water
2–3 carrots, scraped and sliced
2–3 white potatoes, peeled and cut into ½-inch pieces
2–3 stalks of celery, cut into ½-inch slices
2 small yellow squash, sliced
2 small zucchini squash, sliced
1 large onion, chopped
1 small can each: cut green beans, corn kernels, lima beans, and/or garbanzo beans, undrained
2–3 cloves garlic, minced
2 bay leaves
2–3 sprigs parsley, chopped
1 medium-size can whole tomatoes

Bring the stock or water to a boil. If meat is added, cook the meat for 1–1½ hours until very tender, and replace lost water. Add all fresh vegetables plus bay leaf, stir, bring to a boil and simmer for 20–30 minutes. Add all canned vegetables and their juices and simmer 10–20 minutes more. Add parsley just before serving.

Basic Okra Gumbo

Okra gumbo can be made with chicken, wild game, shrimp, crabs, sausage, ham, or whatever is desired.

2 lbs. chicken cut into small pieces, shrimp, or smoked sausage cut into pieces, or 6–8 small crabs (optional)
½–2 cups roux or reduced stock, containing the trinity (chopped onion, celery, and sweet pepper)
2–4 cups okra, diced into ½-inch pieces
1 green onion or shallot finely chopped
2 cloves garlic, minced
3–4 quarts water
2 bay leaves
1 tablespoon Worcestershire sauce
chopped parsley or green onion

If you are using them, brown the chicken pieces first and boil the shrimp or crabs before cleaning and adding. Ham and sausage can be added without prior cooking. Set them aside to prepare the other ingredients.

The diced okra can be fried in a tablespoon of oil for 10–20 minutes to reduce the stickiness of the gums, if desirable. When the stickiness appears to have subsided a little, set this aside.

Heat the roux or reduced stock with the onions, celery and sweet peppers in a large pot until the onions are translucent. Stir in the garlic, green onions and bay leaves, then add the water, Worcestershire sauce, and Tabasco, and bring to a boil. Add the okra and the meat(s) and cook on low heat for 1–2 hours. If no meats or seafood are added, cooking can be cut by ½–1 hour, depending upon the consistency desired. If too thick, more water can be added. Gumbo is served over cooked rice in a soup bowl and garnished with chopped parsley or green onion.

There are as many versions of gumbo as there are Southern chefs. All can be very tasty and nutritious and can include high-fiber foods in addition to, or instead of, okra. Justin Wilson, the noted Cajun chef, prepares a red kidney bean gumbo. My father's favorite (prepared by my grandmothers and mother) was Gumbo Z'Herbes (a gumbo of "herbs", or green leafy vegetables), a very tasty dish that included the chopped leaves of turnip, mustard, and collards; green onions; possibly sweet peppers; and a piece of ham or pickled pork or a ham hock, and served over cooked rice. (This dish could really promote regularity.) Other vegetables have also been used to create nutritious gumbos. One thing that could increase the fiber content of gumbo and add to the flavor, however, would be to serve it over whole brown rice rather than over white milled rice. Brown rice has a mild, nutty flavor when cooked.

Hearty Bean or Lentil Soup

1 lb. dried red kidney beans or lentils (or 2–3 1-lb cans of cooked beans or lentils)
½ tablespoons olive or peanut oil
1 medium onion, peeled and diced
1 medium sweet pepper, diced
2 sprigs parsley, chopped
2 cloves garlic, minced (or ½ teaspoon garlic powder)
2 bay leaves
2–3 quarts water to cover all ingredients in the pot
(optional: a ham hock, pieces of ham or sausage for flavoring)
½ head cabbage, small size, shredded

The dried beans or lentils should be pre-soaked overnight to remove the flatus-causing sugars. Add the oil, onion, sweet pepper, and parsley to a large pot and cook, stirring, until onions are clear. Add garlic and bay leaves, then the water, and bring to a boil. Add the beans or lentils, the ham hock, ham, or sausage, and cook on low heat for 45–60 minutes until the beans feel soft, keeping them covered with water. Remove ¼ to ⅓ of the beans (or lentils) and mash with a fork or in a blender. Return the puree to the pot, stir in the shredded cabbage, and cook 20 minutes more. This can be served over rice or eaten with soup crackers or bread cubes sprinkled on top.

FIBER-RICH VEGETABLE ENTREES

Getting a green and a yellow vegetable and some form of soluble fiber (pectin, gums, mucilages, and soluble hemicelluloses) from vegetables in the principal meal of the day is relatively easy. There are so many to select from as side dishes of steamed, cooked, or raw vegetables that the best recipe is to select those you like, prepare them in a way that you like, and serve them with the main dish.

If you want a completely vegetarian meal, a little more planning will be needed to obtain adequate protein because plant proteins are not as complete as animal proteins. All proteins are made of various amounts and sequences of 22 amino acids, 10 of which are essential for humans; the body cannot synthesize them. They must be obtained from food so that the body can properly resynthesize its tissue proteins. If one amino acid is missing or in short supply, tissue protein synthesis stops, and malnutrition can result. When only plant proteins are consumed, complementary proteins must be eaten to ensure the presence of adequate supplies of all amino acids necessary for proper tissue metabolism. For example, legumes are low in sulfur amino acids, but cereal grains have higher amounts. If legumes (beans or peas) are eaten as a protein source, a cereal grain (wheat or rice) should accompany the legume. In general, beans, peas, lentils, or pulses (legumes) should be accompanied by wheat, rice, barley, corn, or some kinds of nuts or seeds (peanuts, sunflower seeds, cashews, filberts, walnuts, etc.). Adding too many nuts, however, can raise the caloric content of the meal because of the oil contents of these seeds.

Don't forget potatoes as a vegetable. They are versatile and can be steamed, boiled, baked, mashed, or cooked with other vegetables in stews. Dr. Denis Burkitt, the British guru of high fiber, claims that four potatoes a day without sour cream or butter, eaten with the skins, will keep a person "regular" and help control obesity. They have fiber but no fat, and they provide bulk that makes you feel full after a meal so that you do not need sweet desserts to "fill up". A medium-size potato will provide 100–120 calories, about half of the recommended daily allowance (RDA) for vitamin C, plus 5–15% of the niacin, thiamine, and iron needed.

Another way to incorporate extra vegetables is to use pureed vegetables (tomatoes, onions, celery, sweet peppers, carrots, potatoes, mushrooms, and parsley) as a base for preparing thickened gravies or sauces for various meat entrees or side dishes. Mixed vegetable dishes can include just about any that the cook and the family enjoy: those listed plus beans, peas, squashes, green leafy vegetables, all tubers, corn, and eggplants. These can be prepared as vegetable medleys cooked or steamed (excellent for microwave cooking), added to the pans with baked roasts or poultry, or mixed in meat-based stews. The potential for increasing vegetable fiber in the principal meal or entree is limited only by the cook's ingenuity. For this reason, only two recipes are offered here: a vegetable lasagna and New Orleans style red beans and rice.

Eggplant (or Vegetable) Lasagna

6–8 lasagna noodles
1 tablespoon olive or peanut oil
1 cup chopped onions (or 6-8 green onions, chopped)
½ cup parsley, chopped
2 cloves garlic, minced
1 15-oz. can tomatoes, whole, cut into small pieces, plus juice
1 6-oz. can tomato paste
2 cups (14–15 oz.) low-fat ricotta or cottage cheese
1 cup (4 oz.) grated mozzarella cheese
2 medium eggplants, peeled and sliced lengthwise, ½-inch thick (or 2 10-oz. packages frozen chopped spinach, 1 package frozen chopped broccoli, or 5–6 medium carrots, scraped and grated)
½–2 lb. meat, trimmed of all fat and ground (optional)

First, boil the noodles in salted water until soft, drain and set aside. Next, prepare the tomato sauce by heating the oil, onions, parsley, and garlic until onions are clear. If meat is included, add meat and cook until brown. Then add tomato paste, whole tomatoes, oregano, and heat until thickened.

For cheese filling, blend ricotta or cottage cheese in a bowl until smooth. Blanch the eggplant 5 minutes in boiling water and drain. Frozen vegetables should be defrosted and drained. Then in a greased 13″ × 9″ × 2″ baking pan, put a layer of noodles, followed by a layer of eggplant slices (or the other vegetables), spread a layer of the ricotta cheese, followed by the tomato sauce. Repeat the layers noodles, ricotta, vegetables and sauce, and sprinkle grated mozzarella cheese on top. Bake in a 350 °F oven for 40–45 minutes, until cheese on top is brown.

New Orleans Style Red Beans

1 lb. dried red kidney beans
2–3 quarts water
1 ham hock (or 1 lb. ham or cooked sausage, cut in pieces)
1 medium-size green sweet pepper, finely chopped
1 large onion, finely chopped
2 pieces celery with leaves, finely chopped
2 cloves garlic, minced
2 bay leaves
½ cup parsely, finely chopped
cooked rice (if brown rice is used, fiber content is higher)

Presoak beans overnight and discard the water before cooking. Place beans in a large pot and cover with about 2 inches of water. Add the ham hock, ham, or sausage, sweet pepper, onion, celery, garlic, bay leaves, and parsley to the pot and stir. Bring to a boil, then lower the heat and simmer for 1½ hours, stirring periodically, three or four times. When beans become tender, mash some against the sides of the pot to thicken the gravy. Continue heating about ½ hour more, until the beans are very tender. Serve over hot rice, preferably brown rice.

FIBER-RICH DESSERTS

One of the easiest ways to end a meal with fiber-rich foods is the European way—select something from a bowl of fresh fruit. Unfortunately, many people do not think of this as dessert; dessert must be cake, ice cream, or something very sweet. Fruit salad or dried fruit, however, taste sweet and can be an alternate form of fruit to finish a meal. Other types of high-fiber desserts are the high-fiber cookies and muffins, oatmeal cookies with raisins or chopped figs, or cookies containing wheat, oat, or corn bran. Raisins also contain some fiber, plus they have a natural sweetness and chewy texture that can enhance the flavor of most bakery items, whole wheat bread puddings, and whole brown rice puddings. Oatmeal cookies can also serve as a tasty between-meals snack in place of candy bars, ice cream, salted nuts, or deep-fried potato chips or nachos. Popcorn without butter and salt is also a good low-calorie, high-fiber snack. Bran or oatmeal muffins made from whole wheat flour can incorporate blueberries, raisins, or other fruits that increase flavor while providing additional fiber. Most baked goods cookbooks have recipes that can be modified to prepare these cookies and muffins.

Granola bars are somewhat newer products on the market shelves, but they have evolved from low-calorie (110–120 calories) nutrition bars

to confections with enough chocolate and peanut butter to raise the caloric content to 145–150 per bar. Some low-fat granola bars that are available can be consumed as snacks or dessert items after meals without raising the caloric intake excessively for those on weight-control diets. Simply read the labels and avoid those listing chocolate as a major ingredient. Most of the calories in low-fat granola bars are derived from honey, dried fruits, caramel, brown sugar, or corn syrup. These are all natural sources of sugar, but except for dried fruits, they contain no fiber.

However, a simple granola mix that can be prepared at home is tasty, nutritious, and low in fat. Combine one cup each of uncooked rolled oats, raisin bran, and either wheat, corn, or oat bran pulverized to mix well, a half cup each of dried fruits cut into small pieces (dates, figs, prunes, apricots, or more raisins), and unsalted roasted peanut chips, slivered almonds, or sunflower seeds, plus 2–3 tablespoons of honey to hold it together after mixing well. If it is moist enough, the mixture can be pressed into a baking pan and later cut into bars. This home-made granola is a high-fiber product that can also serve as a between-meal snack or as an after-dinner dessert substitute. By adding to rolled oats, cereals selected from Table 6 and dried fruits from Table 7 (based upon personal preference and fiber contents), granola can be readily prepared at home.

I've mentioned citrus peel (albedo) as a high-fiber confection or snack. This was something that Grandma did after fresh grapefruit or oranges were served at breakfast. (In those days very little was wasted or discarded.) Dr. Robert Braddock has done considerable research on citrus peel and pulp at the University of Florida. In 1983 he proposed using washed and dried juice pulp as a water-binding agent or thickener in gravies and sauces; it was shown to absorb 10 times its weight of water. This product was 19% crude fiber, 21% protein, and 36% carbohydrate. Other proposed uses were as hydrophilic ingredients in high-fiber bread and cakes. Dried citrus peel albedo had 11.2% dietary fiber (compared to 3.7% for apple peel and 8.6% for pear peel) and could be used in applications similar to those for the pulp. Grandma's recipe for a citrus peel confection follows.

Candied Citrus Peel

6 large oranges or 3–4 grapefruit
2–4 cups sugar (depending upon personal taste)
lots of water

Using a sharp knife, cut the peel into ½–1 inch strips, remove the adhering pulp, and (optional) scrape away or peel the thin outer colored layer of the peel. This outer layer contains oil droplets that some people consider bitter-tasting. The albedo also contains bitter compounds, but these are water-soluble. If time is not a priority, the strips of peel can be covered with water, stored 6–8 hours or overnight in a refrigerator, then removed and squeezed by hand; the water is discarded, and the peels are covered again with water. The leaching process is repeated three or four times until the bitterness is removed. If this is too long, the strips can be covered with water, boiled 5 minutes, drained, rinsed with cold water, and boiled again three or four times until the bitterness is removed. Strips should be soft but not mushy.

When debittering is completed, place the sugar into 1½ cups of water in a pot and bring to a boil to produce a clear syrup. Lower the heat, add the strips of peel and cook over low heat for 15–20 minutes, stirring, until the strips are translucent. If necessary, add a little water to avoid dryness during cooking. Remove the strips from the pot and allow to drain dry on a cookie sheet overnight. Candied peel can be stored in airtight containers when dry. It can be consumed as is as a snack, added to cakes and muffins like dried fruit, cut into smaller pieces for use as a garnish on fruit salad, or coated with semi-sweet chocolate and eaten as a fiber confection. (For this, melt some semi-sweet chocolate in a pan, roll the strips in the melted chocolate, and place on a cookie sheet until dry.)

Another dessert that can add ample fiber to a meal is New Orleans style bread pudding.

Bread Pudding

½–¾ loaf of stale (or dried) bread, preferably whole wheat, broken into pieces
3–4 cups skim milk
1 cup firmly packed brown sugar
7–8 tablespoons margarine, melted
2 eggs
1 cup raisins
1 16-oz. can sliced peaches, drained and cut into pieces
1 medium size can pears, drained and cut into pieces
1 cup shredded coconut

½ cup maraschino cherries, cut into pieces
2 tablespoons vanilla
1 teaspoon ground cinnamon
½ teaspoon grated nutmeg

Mix all ingredients in a large bowl to moisten bread well (but not to a soupy consistency). Pour into a greased 3-quart baking dish and bake in a 360 °F oven for 1½ hours, until top is golden brown. In New Orleans, bread pudding is served warm with a whiskey or rum sauce topping, but this is not critical. The flavor is excellent by itself.

CONCLUSIONS

We hear so much today about increasing fiber consumption; cutting down on fat, sugar, and salt; and increasing exercise. Does this mean we should avoid cocktail parties unless we bring along our own bran muffins or a bag of hay? Not unless we also bring a small saddle horse or pony to the party. (This is not recommended unless the host or hostess has an extremely good sense of humor.) But this extreme measure is, understandably, not necessary if we always remember: BALANCE AND MODERATION. If a person is obese or has high blood pressure, high blood levels of cholesterol or triglycerides, heart disease, or any disorder that can be affected by rich, high-calorie foods, it is not just a matter of adding dietary fiber but a matter of avoiding some foods to achieve good health. For the general party-goer, however, it's simply a matter of not gorging on the richest snacks, all-meat or cheese items, but to balance these with raw vegetables and fruit snacks and whole wheat crackers and breads.

Even the fiber advocates warn against excessively high amounts of fiber and sudden major changes in the diet because the body could become mineral deficient or suffer GI disturbances. The digestive system takes a while to adapt to drastic changes in the diet, so changes should be gradual. Also, scientists still have not determined the proper amount of fiber needed because the sources and types of fiber vary widely in foods. Nutritionists do recommend fiber as part of a properly balanced meal and suggest 20–30 grams per day from whole cereal grains, fruits, and vegetables.

Our knowledge of fiber in the diet and its effects on health care is still incomplete, but it is extensive. When Grandma said that a balanced meal should contain a protein, a starch, a green and a yellow vegetable, and roughage, she didn't know the biochemical basis for this description of a balanced meal. But modern research has shown that she was on the right track. Today we know which forms of fiber promote laxation and bowel motility, which forms hold large amounts of water, and which ones can bind bile acids and affect cholesterol levels in the blood. It is simply a matter of selecting foods wisely, eating wisely, and exercising will-power at buffets and smorgasbords. As research continues to identify new roles for dietary fiber, food manufacturers will continue to produce more flavorful fiber products, such as ready-to-eat breakfast cereals, as well as the traditional low-fiber food products now supplemented with a compatible form of fiber.

Bibliography

Dietary Fiber in Foods; Bulletin A598; Purdue-Frederick: Norwalk, CT, 1982.

"Research Update: Protein and Cancer Risk"; *Am. Inst. Cancer Res. Newsl.* **1986,** *12,* 12.

Buyer's Guide to Dietary Fiber; California Fig Advisory Board Bulletin, Fresno, CA, 1987.

Adair, D. *Selling Nutrition; Cereal Foods World* **1986,** *31(5),* 355.

Al-Wandawi, H. "Chemical Composition of Seeds of Two Okra Cultivars," *J. Agric. Food Chem.* **1983,** *31,* 1355.

Anderson, J. W. "Physiological and Metabolic Effects of Dietary Fiber," *Fed. Proc.,* **1985,** *44,* 2902.

Anderson, J. W.; Tietzen-Clark, J. "Nutrition Management of Hyperlipidemia and Diabetes," *Food Nutr. News* **1987,** *59 (4),* 67.

Baker, D.; Holden, J. M. "Fiber in Breakfast Cereals," *J. Food Sci.* **1981,** *46,* 396.

Braddock, R. J. "Utilization of Citrus Juice Vesicle and Peel Fiber," *Food Technol.* **1983,** *37 (12),* 85.

Braddock, R. J.; Crandell, P. G. "Carbohydrate Fiber from Orange Albedo," *J. Food Sci.* **1981,** *46,* 650.

Black, H. S. "Utility of the Skin/UV-Carcinogenesis Model for Evaluating the Role of Nutritional Lipids in Cancer," In *Diet, Nutrition and Cancer: From Basic Research to Policy Implication;* Alan R. Liss: New York, 1983, p 49.

Black, H. S. "Photocarcinogenesis and Diet," *Fed. Proc.,* **1987,** *46,* 1901.

Bond, J. H.; Levitt, M. D. "Effect of Dietary Fiber on Intestinal Gas Production and Small Bowel Transit Time in Man," *Am. J. Clin. Nutr.* **1978,** *31,* S169.

Caldwell, E. F. "How Can We Lose?" *Cereal Foods World* **1988,** *33 (3),* 264.

Camire, A. L.; Clydesdale, F. M. "Interactions of Soluble Iron with Wheat Bran," *J. Food Sci.* **1982,** *47,* 1296.

Carroll, K. K. "Hypercholesterolemia and Atherosclerosis: Effects of Dietary Protein," *Fed. Proc.,* **1982,** *41,* 2792.

Catignani, G. L.; Carter, M. E. "Antioxidant Properties of Lignin," *J. Food Sci.* **1982,** *47,* 1745.

Chen, J. Y.; Piva, M.; Labuza, T. P. "Evaluation of Water-Binding Capacity of Food Fiber Sources," *J. Food Sci.* **1984,** *49,* 59.

Chou, M. "How We Got to Where We Are," *Cereal Foods World* **1988,** *33 (5),* 456.

Cummings, J. H.; Branch, W.; Jenkins, D. J. A.; Southgate, D. A. T.; Houston, H.; James, W. P. T. "Colonic Response to Dietary Fibre from Carrot, Cabbage, Apple, Bran, and Guar Gum," *The Lancet* **1978,** *I,* 5.

DeGroot, A. P.; Luyken, R.; Pikaar, N. A. "Cholesterol-Lowering Effect of Rolled Oats," *The Lancet* **1963,** *2,* 303.

DeLucca, II, A. J.; Plating, S.J.; Ory, R. L. "Isolation and Identification of Lipolytic Microorganisms Found on Rough Rice from Two Growing Areas," *J. Food Protect.* **1978,** *41,* 28.

Dintzis, F. R.; Watson, P. R. "Iron Binding of Wheat Bran at Human Gastric pH," *J. Agric. Food Chem.* **1984,** *32,* 1331.

Eastwood, M. A.; Smith, A. N.; Mitchell, W. D.; Pritchard, J. L. "Physical Characteristics of Fiber Influencing the Bowel," *Cereal Foods World* **1977,** *22 (1),* 10.

Ellis, J. "The Laxative Industry Has America in a Bind," *Moneysworth* **1976,** *Nov. 22,* 14.

Erickson, K. L. "Dietary Fat and Tumorigenesis in Laboratory Animals," *Food Nutr. News* **1984,** *56 (2),* 9.

Fast, R. B. "Breakfast Cereals: Processed Grains for Human Consumption," *Cereal Foods World* **1987,** *32 (3),* 241.

Gormley, T. R. "Fibre in the Diet," *Farm Food Res.* **1977,** *8 (4),* 99.

Hardin, B. "Grains Hold the Key to Reducing Blood Cholesterol," *Agric. Res.* **1985,** *33 (91),* 10.

Hashimoto, S.; Shogren, M. D.; Bolte, L. C.; Pomeranz, Y. "Cereal Pentosans: Their Estimation and Significance. III. Pentosans in Abraded Grains and Milling By-Products," *Cereal Chem.* **1987,** *64,* 39.

Heller, S. N.; Hackler, L. R. "Water-Holding Capacity of Various Sources of Plant Fiber," *J. Food Sci.* **1977,** *42,* 1137.

Herranz, J.; Vidal-Valverde, C.; Rojas-Hidalgo, E. "Cellulose, Hemicellulose, and Lignin Content of Raw and Cooked Processed Vegetables," *J. Food Sci.* **1983,** *48,* 274.

Hoo, A. F.; McLellan, M. R. "The Contributing Effect of Apple Pectin to the Freezing Point Depression of Apple Juice Concentrates," *J. Food Sci.* **1987,** *52,* 372.

Hughes, R. E. "Hypothesis: A New Look at Dietary Fibre," *Human Nutr.: Clin. Nutr.* **1986,** *40 C,* 81.

Karlstrom, B.; Vessby, B.; Asp, N-G.; Boberg, M.; Gustafsson, I.-B.; Lithell, A.; Werner, I. "Effects of an Increased Content of Cereal Fibre in the Diet of Type 2 (Non-Insulin Dependent) Diabetic Patients," *Diabetologia* **1984,** *26,* 272.

Kelsay, J. L. "Effect of Diet Fiber Level on Bowel Function and Trace Mineral Balances of Human Subjects," *Cereal Chem.* **1981,** *58,* 2.

Kelsay, J. L.; Jacob, R. A.; Prather, E. A. "Effect of Fiber from Fruits and Vegetables on Metabolic Responses of Human Subjects. III. Zinc, Copper, and Phosphorus Balances," *Am. J. Clin. Nutr.* **1979,** *32,* 2307.

Kern, F., Jr.; Birkner, H. J.; Ostrower, V. S. "Binding of Bile Acids by Dietary Fiber," *Am. J. Clin. Nutr.* **1978,** *31,* S175.

Kies, C.; Beshgetoor, D.; Fox, H. M. "Dietary Fiber and Zinc Bioavailability for Humans," In *Antinutrients and Natural Toxicants in Foods;* Ory, R. L., Ed.; Food and Nutrition Press: Westport, CT, 1981, p 319.

Kolata, G. "Dietary Dogma Disproved," *Science* **1983,** *220,* 487.

Klurfeld, D. M.; Kritchevsky, D. "Influence of Animal and Vegetable Protein on Serum Cholestrol, Lipoproteins, and Experimental Atherosclerosis," In *Plant Proteins: Applications, Biological Effects, and Chemistry;* Ory, R. L., Ed.; ACS Symposium Series 312; American Chemical Society: Washington, DC, 1986, p 150.

Kritchevsky, D. "Influence of Dietary Fiber on Bile Acid Metabolism," *Lipids* **1987,** *13,* 982.

Kritchevsky, D. "Diet and Atherosclerosis: Everything Counts," *Cereal Foods World* **1983,** *28 (7),* 415.

Kritchevsky, D. "The Role of Calories and Energy Expenditure in Carcinogenesis," *Food Nutr. News* **1987,** *59 (3),* 63.

LaBell, F. "Powdered Cellulose Adds Texture, High Fiber to Reduced-Calorie Foods," *Food Process.* **1896,** *47 (7),* 45.

Marlett, J. A.; Chesters, J. G. "Measuring Dietary Fiber in Human Foods," *J. Food Sci.* **1985,** *50,* 410.

Miller, W. B. "Seeing the 'Light': Cellulose in Reduced-Calorie Baked Goods," *Technical Bulletin;* James River Corp, July 1986.

Mod, R. R.; Conkerton, E. J.; Ory, R. L.; Normand, F. L. "Hemicellulose Composition of Dietary Fiber of Milled Rice and Rice Bran," *J. Agric. Food Chem.* **1978,** *26,* 1031.

Mod, R. R.; Conkerton, E. J.; Ory, R. L.; Normand, F. L. "Composition of Water-Soluble Hemicelluloses in Rice Bran from Four Growing Areas," *Cereal Chem.* **1979,** *56,* 356.

Mod, R. R.; Ory, R. L.; Morris, N. M.; Normand, F. L. "Chemical Properties and Interactions of Rice Hemicellulose with Trace Minerals in Vitro," *J. Agric. Food Chem.* **1981,** *29,* 449.

Mod, R. R.; Ory, R. L.; Morris, N. M.; Normand, F. L. "In Vitro Interaction of Rice Hemicellulose with Trace Minerals and Their Release by Digestive Enzymes," *Cereal Chem.* **1982,** *59,* 538.

Mod, R. R.; Ory, R. L.; Morris, N. M.; Normand, F. L.; Saunders, R. M.; Gumbmann, M. R. "Effect of Rice Hemicellulose and Microcrystalline Alpha-Cellulose on Selected Minerals in the Blood and Feces of Rats," *J. Cereal Sci.* **1985,** *3,* 87.

Monte, W. C.; Maga, J. A. "Extraction and Isolation of Soluble and Insoluble Fiber Fractions from the Pinto Bean (*Ph. vulgaris*)," *J. Agric. Food Chem.* **1980,** *28,* 1169.

Normand, F. L.; Ory, R. L.; Mod, R. R. "In Vitro Binding of Bile Acids by Rice Hemicelluloses," In *Dietary Fibers: Chemistry and Nutrition;* Inglett, G. E.; Falkehag, S. I., Eds.; Academic: New York, 1979, p 203.

Normand, F. L.; Ory, R. L.; Mod, R. R. "Interactions of Several Bile Acids with Hemicelluloses from Several Varieties of Rice," *J. Food Sci.* **1981,** *46,* 1159.

Normand, F. L.; Ory, R. L. "Effect of Rice Hemicelluloses on Pancreatic Lipase Activity in Vitro," *J. Food Sci.* **1984,** *49,* 956.

Normand, F. L.; Ory, R. L.; Mod, R. R.; Saunders, R. M.; Gumbmann, M. R. "Influence of Rice Hemicellulose and Alpha-Cellulose on Lipid and Water Content of Rat Feces and on Blood Lipids," *J. Cereal Sci.* **1984,** *2,* 37.

Normand, F. L.; Ory, R. L.; Mod, R. R. "Binding of Bile Acids and Trace Minerals by Soluble Hemicelluloses of Rice," *Food Technol.* **1987,** *41 (2),* 86.

Olson, A.; Gray, G. M.; Chiu, M.-C. "Chemistry and Analysis of Soluble Dietary Fiber," *Food Technol.* **1987,** *41 (2),* 71.

Ory, R. L.; Bog-Hansen, T. C.; Mod, R. R. "Properties of Hemagglutinins in Rice and Other Cereal Grains," In *Antinutrients and Natural Toxicants in Foods;* Ory; R. L., Ed.; Food Nutrition Press: Westport, CT, 1981, p 159.

Ory, R. L.; Flick, G. J., Jr. "Peanut Skins: A Potential Source of Dietary Fiber for Foods," *Proc. Am. Peanut Res. Educ. Soc.* **1987,** *19,* 44.

Owen, D. F.; Cotton, R. H. "Dietary Fibers," *Cereal Foods World* **1982,** *27 (10),* 519.

Pariza, M. W. "Analyzing Current Recommendations on Diet, Nutrition, and Cancer," *Food Nutr. News* **1986,** *58 (1),* 1.

Payne, T. J. "The Role of Raisins in High-Fiber Muesli-Style Formulations," *Cereal Foods World* **1987,** *32 (8),* 545.

Phillips, R. L.; Snowden, D. A. "Mortality among Seventh-Day Adventists in Relation to Dietary Habits and Life Style," In *Plant Proteins: Applications, Biological Effects, and Chemistry;* Ory, R. L., Ed.; ACS Symposium Series 312; American Chemical Society: Washington, DC, 1986, p 162.

Plat, D.; Ben-Shalom, N.; Levi, A.; Reid, D.; Goldschmidt, E. E. "Degradation of Pectic Substances in Carrots by Heat Treatment," *J. Agric. Food Chem.* **1988,** *36,* 362.

Rose, H. E.; Quarterman, J. "Dietary Fibers and Heavy Metal Retention in the Rat," *Environ. Res.* **1987,** *42,* 166.

Rydning, A.; Brestad, A. "Dietary Fiber and Peptic Ulcer," *Scand. J. Gastroentr.* **1986,** *21 (1),* 1.

Salyers, A. A.; Palmer, J. K.; Wilkins, T. D. "Degradation of Polysaccharides by Intestinal Bacterial Enzymes," *Am. J. Clin. Nutr.* **1978,** *31,* S128.

Salyers, A. A.; Palmer, J. K.; Balascio, J. "Digestion of Plant Cell Wall Polysaccharides by Bacteria from the Human Colon," In *Dietary Fibers: Chemistry and Nutrition;* Inglett, G. E.; Falkehag, S. I., Eds.; Academic: New York, 1979, p 193.

Scala, J. "Fiber: The Forgotten Nutrient," In *The Nutrition Crisis: A Reader;* Labuza, T. P., Ed.; West: St. Paul, MN, 1975, p 52.

Schweizer, T. F. "Dietary Fibre in the Swiss Diet," *Nestle Research News 1986/87;* Nestle Ltd.: Vevey, Switzerland, 1987, p 200.

Seibert, S. E. "Oat Bran as a Source of Soluble Dietary Fiber," *Cereal Foods World* **1987,** *32 (8),* 552.

Somogyi, L. P. "Prunes, a Fiber-Rich Ingredient," *Cereal Foods World* **1987,** *32 (8),* 541.

Spiller, G. A. *CRC Handbook of Dietary Fiber in Human Nutrition;* CRC Press: Boca Raton, FL, 1986, p 483.

Steiner, G. "Diabetes and Atherosclerosis," *Diabetes* **1981,** *30 (Suppl. 2),* 1.

Stephen, A. M.; Cummings, J. H. "Water-Holding by Dietary Fibre in Vitro and its Relationship to Faecal Output in Man," *Gut* **1979,** *20,* 722.

Story, J. A.; Kritchevsky, D. "Dietary Fiber and Lipid Metabolism," In *Fiber in Human Nutrition;* Spiller, G. A.; Amen, R. J., Eds.; Plenum: New York, 1976, p 171.

Story, J. A.; Kritchevsky, D. "Bile Acid Metabolism and Fiber," *Am. J. Clin. Nutr.* **1978,** *31,* S199.

Toma, R. B.; Curtis, D. J. "Dietary Fiber: Effect on Mineral Bioavailability," *Food Technol.* **1986,** *40 (2),* 111.

Vahouny, G. V. "Dietary Fiber, Lipid Metabolism, and Atherosclerosis," *Fed. Proc.,* **1982,** *41,* 2801.

Walter, R. H.; Rao, M. A.; Sherman, R. M.; Cooley, H. J. "Edible Fibers From Apple Pomace," *J. Food Sci.* **1985,** *50,* 747. Additional Technical Reading

Amado, R.; Schweizer, T. F., Eds.; *Nahrungsfasern—Dietary Fibers;* Academic: London, 1986.

Bjoerntorp, P.; Vahouny, G. V.; Kritchevsky, D., Eds.; *Current Topics in Nutrition and Disease;* Vol. 14. *Dietary Fiber and Obesity;* Alan R. Liss: New York, 1985.

Dreher, M. L. *Handbook of Dietary Fiber: An Applied Approach;* Marcel Dekker: New York, 1987.

Fishman, M. L.; Jen, J. J., Eds.; *Chemistry and Function of Pectins;* ACS Symposium Series 310; American Chemical Society: Washington, DC, 1986.

Furda, I., Ed.; *Unconventional Sources of Dietary Fiber: Physiological and In Vitro Functional Properties;* ACS Symposium Series 214; American Chemical Society: Washington, DC, 1983.

Heaton, K. W. *Dietary Fiber—Current Developments of Importance to Health;* Food and Nutrition Press: Westport, CT, 1979.

Leeds, A. R., Ed.; *Dietary Fiber Perspectives;* Food and Nutrition Press: Westport, CT, 1985.

Spiller, G. A.; Amen, R. J., Eds. *Topics in Dietary Fiber Research;* Plenum: New York, 1978.

Trowell, H. "Definition of Dietary Fiber and Hypotheses That It Is a Protective Factor in Certain Diseases," *Am. J. Clin. Nutr.* **1976** , *29* , **417.**

Trowell, H.; Burkitt, D.; Heaton, K., Eds. *Dietary Fibre, Fibre-Depleted Foods and Disease;* Academic: London, 1985.

Index

A

Abdominal bloating, 57
Abdominal pain, from fiber, 104
Acid detergent fiber (ADF) method, 21
Adolescents, food choices, 129–131
Adults, food choices, 129–131
Adventist lifestyle, health benefits, 68
Aerobic exercise, health benefits, 41
African diet, relation to cancer, 4
Agar, 19,84
Age, cancer risk factor, 66
Aging
  effect on nutrient requirements, 62
  nutritional problems, 49–50
Alfalfa, interaction with bile acids, 31
Alfalfa seeds, source of mucilage, 20
Alginates, 19, 58
Alginic acid, binding of heavy metals, 84
All-Bran, 55,101,102
All-Bran Extra Fiber, 102
Almonds, fiber content, 121
Alpha-cellulose (Alphacel)
  effect on fecal water content, 52
  use as food additive, 13,35
American Cancer Society, 65,112
American Council on Science and Health, 112
American Heart Association, 27,41,45,108
American Institute for Cancer Research, 68,112
Amylase, 12,48
Anemia, prevention, 76
Animal fats, relation to cancer, 62,66
Animal proteins
  effect on cholesterol levels, 30
  negative health benefits, 68
Anti-cancer foods, 64
Anti-sterility vitamin, 93
Antioxidants, health benefits, 16,93
Appendicitis
  effect of diet, 4,6
  effect of laxative use, 51
Apple
  fiber content, 114
  pectin content, 17
  water-holding capacity, 55
Apple pomace
  fiber content, 113
  pectin content, 17
  water-holding capacity, 55
Apricot, fiber content, 114
Arabinose, 15,18,82
Atherogenicity, 29
Atherosclerosis
  effect of diabetes, 69
  relation to diet, 6,28
Atherosclerosis and cholesterol, 28–30

B

B vitamins, function, 92
Bad cholesterol, negative health effects, 27
Balanced meal
  composition, 3
  importance, 111–112,147
Ballpark peanuts, 121
Banana
  fiber content, 114
  water-holding capacity, 55
Barley
  health benefits for diabetics, 70
  source of dietary fiber, 107
Barley bran, fiber content, 106
Battle of the Brans, 99–109
Beans
  effect of cooking, 112
  fiber content, 118
  source of gums, 19

Beets, fiber content, 118
Belching, cause, 58
Beta-glucan, 108
Bible, advocacy of food fiber, 111
Bile, release from gallbladder, 30
Bile acid(s)
  binding by dietary fibers, 29
  binding by triglycerides, 30–33
  conversion from cholesterol, 26
  production of carcinogens, 64
  role in digestion, 30,48
Bile acid excretion
  cholesterol elimination, 29
  effect of fiber, 29
Bioavailability of trace minerals, 75
Blackberry, fiber content, 114
Blood coagulation factor in whole brown rice, 81
Blood sugar, effect of starchy foods, 71
Blueberry, fiber content, 114
Bowel functions, effect of fiber, 4,47,49,51
Bowel regularity, effect of diet, 5,115–117
Bran
  addition to foods, 126–127
  effect on bowels, 4
  effect on precancerous polyps, 65
  health benefits, 3
  source of food-grade cellulose, 13
Bran Muffin Crisp, 102
Brazil nuts, fiber content, 121
Bread, effect on serum sugar level, 71
Bread–bean combination, effect on serum sugar level, 71
Bread pudding, recipe, 145–146
Breakfast, importance, 45
Breakfast cereals
  advertising claims, 101
  fiber content, 101
  neutral detergent fiber, 103–104
  nutritional data, 102–103
  salt content, 101
  source of fiber, 39
  sugar content, 102
Breakfast cereals industry, 99
Broad beans, fiber content, 118
Broccoli
  anti-cancer food, 65
  fiber content, 118
Brown bread, effect on iron balance, 83
Brussels sprouts
  anti-cancer food, 65
  fiber content, 118
Burkitt, Dennis, 4–6,47,62
Burkitt's lymphoma, 4

C

Cabbage
  anti-cancer food, 65
  fiber content, 118
  source of oxalic acid, 77
Cadmium, binding by pectins, 84
Calcium
  binding by hemicellulose, 77
  binding by oxalic acid, 77
  binding by phytic acid, 75,85
  effect of wheat bran on absorption, 82
  function, 76,93,126
Calcium oxalate, formation, 77
Cancer
  public fear of, 62
  relation to diet, 4,66
Cancer-preventing compounds, 64
Candied citrus peel, recipe, 144–145
Canola oil, 28
Carbohydrates
  caloric contribution to diet, 89
  functions, 91
  metabolism, 96
  sources, 91
  sugar polymers, 11
Carboxymethylcellulose, cellulose derivative, 13,94
Carcinogens
  effect of contact with colon wall, 63
  production, 4
Carotenoids, cancer-preventing compound, 64
Carrageenans, 19,55,84
Carrots
  effect of cooking, 112
  fiber content, 118
  source of phytic acid, 85
  water-holding capacity, 55
Carver, George Washington, 121
Cauliflower
  anti-cancer food, 65
  water-holding capacity, 55
Celery
  fiber content, 118
  source of food-grade cellulose, 13
Celiac disease, 107
Cellophane, interaction with bile acids, 31
Cellulose
  binding of heavy metals, 84
  degradation during digestion, 56
  dietary fiber, 8,12–14
  effect on fecal water content, 52
  effect on mineral balances, 82–83
  interaction with bile acids, 31

Cellulose–*Continued*
use as food additive, 35
water-holding capacity, 55
Cellulose derivatives, 13,14
Cereal brans
degradation, 56,58
effect on stool weight and colonic motility, 54
pectin content, 17
source of phytic acid, 85
water-holding capacity, 54
Cereal fiber, effect on precancerous polyps, 65
Cereal grains, basic structure, 105
Cereal industry, 99
Cereal starches, effect on serum sugar level, 71
Cheerios, 108
Cheese–bread combination, effect on serum sugar level, 71
Chemical analysis of fiber in food, 20–22
Chemical antioxidants, for storage of fruits, 116
Chemical bond, 11
Cherry, fiber content, 114
Children, food choices, 129–131
Cholesterol
effect of dietary fiber, 29–30
effect of dietary proteins, 30
factors affecting levels in the blood, 30
good and bad forms, 26
major component of plaques, 28
relation to atherosclerosis, 28–30
structure, 26
synthesis in liver, 26
Cholic acid, binding by rice hemicelluloses, 32
Chymotrypsin, role in digestion, 48
Circulatory disorders, effect of diet, 6
Citrus, pectin content, 17
Citrus peel, pectin content, 17
Citrus peel albedo, uses, 144
Clostridia (colon bacteria), production of carcinogens, 64
Clusters, 102
Coconut, fiber content, 114
Colon bacteria, activity, 56–57
Colon cancer
effect of diet, 4,6,47
risk factors, 66,68–69
Colon diseases, effect of fiber, 47
Colon–rectal cancer
effect of diet, 4,62–69
factors, 63–64
Commercial pectin, 17
Common Sense Oat Bran, 108
Complex carbohydrates, relation to colon–rectal cancer, 62
Constipation
effect of diet, 5,47–59
prevention, 59,104
treatment with fiber, 96
Cooking, effect on dietary fiber, 112
Copper
binding by hemicellulose, 77
binding by protein, 81
effect of pectins, 83
function, 76,93
release from bound state by enzymes, 80
Copper balance, effect of diet, 83
Corn
effect on serum sugar level, 71
effective binder of bile acids, 31
fiber content, 118
Corn bran
effect on mineral balance, 83
vs. wheat bran, 101,105
water-holding capacity, 55
Coronary disease, effect of animal protein, 68
Cotton, 12
Cracker Jack, 121
Cracklin' Oat Bran, 102
Crapo, Phyllis, 70
Creole cooking, 137
Creole okra gumbo, 120
Crude fiber, 11,21,94
Crude fiber (CF) method, 21

D

Dairy products
effect on serum sugar level, 71
importance, 126
Daniel, advocacy of food fiber, 111
Date, fiber content, 114
Deaton, John, 51
DeCosse, Jerome, 65
Del Monte Corporation, 89
Dental plaque, relation to fiber, 73
Department of Health and Human Services, 99,112
Designer foods, 64
Desserts high in fiber, 142–146
Diabetes
appropriate diet, 70–73
deaths caused by, 69
effect of animal protein, 68
Diabetes mellitus, 70
Dialysis, apparatus, 78
Diet
cancer risk factor, 66
effect on serum sugar level, 71
relation to atherosclerosis, 28

Dietary fat
digestion, 26
effect on cholesterol levels, 26
relation to cancer, 66
Dietary fiber
beneficial effects, 59,73
Burkitt's work, 4–6
complexity of analysis, 20–22
definitions, 3
effect on bowel movement, 49
effect on colon bacteria, 57
effect on feces, 4,56
effect on human health, 28,90
essentiality, 95–97
health benefits, 5,33,47
major types, 8
natural laxative, 51–56,96
pioneers in research, 6
relation to cancer, 4,61–69
role in controlling obesity, 35
sources, 3,8–9,91,99
theory of effect on digestive diseases, 6
variation in composition and properties, 9
variation in reactions and health benefits, 9
water-retaining ability, 47
See also Fiber
Dietary fiber complex, 94
Dietary fiber recipes, 131–146
Dietary guidelines, 40
Dietary habits, effect on health, 68
Dietary protein, effect on cholesterol levels, 30
Dieting, role in weight control, 38
Digestion, 48–49
Digestive diseases, effect of diet, 6
Diverticular disease
effect of diet, 6
prevention by dietary fiber, 59
relation to diet, 47
result of constipation, 51
Diverticular pockets, 58
Diverticulitis, effect of prolonged laxative use, 51
Diverticulosis, relation to diet, 4
Dough fermentation, release of zinc, 85
Dried beans
nutritional benefits, 119,130
remedy for gas-forming sugars, 130
source of phytic acid, 84,85
Dried fruits, fiber content, 114
Dried prunes
method for softening, 116
promotion of bowel regularity, 115
Drug therapy, treatment of diabetes, 69
Duodenal ulcers, effect of diet, 73

E

Eastwood, Martin, 6,53
Eating habits
changes, 123
obstacles to new habits, 125
role of proper education, 125
Eating wisely, importance, 147
Economic status, relation to colon–rectal cancer, 63
Edible fiber, 94
Eggplant
fiber content, 118
healthy cooking method, 120
Elderly, nutrition problem, 124
Erickson, Kent, 66
Essential nutrients, 89
Excessive weight, cancer risk factor, 66
Exercise
calories used in activities, 44
health benefits, 41
importance, 147
relation to cancer, 69
role in weight control, 41–44

F

Family history, cancer risk factor, 66
Fast food(s), 130
Fast-food meals, characteristics, 90
Fat
adverse effects, 92
caloric contribution to diet, 89
connotations, 25
metabolism, 96
recommended intake, 97
Fat absorption, 37–38
Fat depots, adipose tissue, 96
Fat excretion
effect of fiber, 36–38
effect of pancreatic lipase activity, 37–38
effect on body weight and waistline, 37
Fecal bulk, effect of fibers, 56,59
Fecal discharges, effect of diet, 4
Fiber
addition to foods, 126–127
beneficial effects, 109
chemical analysis, 20–22
chemical forms, 11
definition, 7
degradation in colon, 58
effect of cooking, 112
effect on fat excretion, 36–38
effect on fecal water content, 52

Fiber–*Continued*
effect on pancreatic lipase activity, 37
interaction with minerals, 77
passage through GI tract, 7–8
peptic ulcers, 73
problem caused by, 95
recommended daily intake, 94
role in cancer prevention, 69
sources, 11
variation in cereal grains, 106
variation in types and amounts, 8–9
variation in vegetables, 118–119
what it is, 11–22
where to find it, 11–22
See also Dietary fiber
Fiber and mineral nutrition, 75–86
Fiber–bile acid interactions, 31
Fiber-containing foods, water-holding capacity, 55
Fiber deficiency, treatment, 94
Fiber One, 101,102
Fibermania, 3–9
Fiber-rich desserts, 142–146
Fiber-rich vegetable entrees, recipes 140–142
Figs
fiber content, 114,116
natural humectant, 116
promotion of bowel regularity, 116
First Health and Nutrition Examination Survey, 65
Fish oil, health benefits, 27,92
Flavonoids, cancer-preventing compounds, 64
Food additives
good chemicals, 19–20
negative public perception, 19
Food advertising, 89
Food analysis, variability of results, 54
Food and Drug Administration, 19,93
Food fibers, examples, 7
Food-grade (powdered) cellulose, 13
Food gums
use as food additives, 20
water-holding capacity, 55
See also Gum(s)
Food labels, importance, 125,126
Food processing, effect on dietary fiber, 112,117
Food products, nutritional labeling, 89
Food-related surveys, 65
Fresh fruits
effect on iron balance, 83
water-holding capacity, 54
Frozen juice concentrates, effect of pectin, 18
Fruit(s)
fiber content, 113–117
Fruit(s)–*Continued*
health benefits due to fiber, 111–121
health benefits for diabetics, 70
raw vs. processed, 113,114–115
sulfite treatment, 116
types of dietary fiber, 112
Fruit pectins, properties, 17
Fruit Wheats, 103

G

Galactomannans, neutral mucilage, 20
Galactose, 15,18
Galacturonic acids, 15,17
Gallbladder disease, 69
Gallstones, 26,69
Gas-forming sugars in dried beans, removal, 130
Gas production, GI tract, 57–58
Gastrointestinal (GI) tract
gas production, 57–58
effect of diet on diseases, 5
Gel-forming additives, water-holding capacity, 55
General Mills, 101,102
Geriatric cocktail, 116
Glucose, 12,15
Glucuronic acids, 15
Gluten, 107
Glycerides, 26
Glycocholic acid, 32
Glycogen, 48,96
Glycotaurocholic acid, 32
Good cholesterol, health benefits, 26
Good health
role of proper nutrition, 124
steps, 124–126
Graham, Sylvester, 4
Graham cracker, 4
Graham flour, 4
Granola mix, preparation, 143–144
Grape, fiber content, 115
Green beans, fiber content, 118
Green leafy vegetables, source of oxalic acid, 77
Guar, source of mucilage, 20
Guar gum, 19, 58
Gum(s)
components, 18
degradation during digestion, 56
dietary fiber, 8
effect on bile acid excretion, 29
effect on serum cholesterol, 29
sources, 19
types, 18, 19
See also Food gums

Gum arabic, 18, 58
Gum disease, relation to fiber, 73
Gum ghatti, 18, 58
Gum tragacanth, 19, 58
Gumbo(s)
basic okra gumbo, 138
gumbo Z'Herbes, 139
okra gumbo, recipe, 138
origin of name, 136
red kidney bean gumbo, 139

H

Health benefits of exercise, 41
Health benefits of fiber
antioxidants, 14
gums, 19
hemicellulose, 14
lignin, 16
pectin, 17
to African natives, 4–6
Healthy meals
increasing fiber content, 126–127
planning, 125–126
Heart disease
lifestyle factors, 42
major cause, 28
relation to cholesterol, 28–29
Heart rates during exercise, 42
Heaton, Kenneth, 4
Heavy metals, binding by pectins, 84
Hemicellulase, release of bound minerals, 80
Hemicelluloses
binding of minerals, 77
degradation, 56,58
dietary fiber, 8
effect on mineral balances, 82,83
human health benefits, 14–16
Hemoglobin, synthesis, 76
Hemophilia, 93
Hemorrhoids, effect of diet, 4,6,47,48
Hiatal hernia, effect of diet, 6,48
High-density lipoprotein, good cholesterol, 26
High-fat diets, relation to cancer, 66
High-fiber bran cereals, 101
High-fiber cereals
fiber content, 101
nutritional content, 102–103
High-fiber diets
effect on bile acid excretion, 29
effect on carcinogens, 63
effect on minerals, 75,76,84
health benefits for diabetics, 70,72
need for mineral supplementation, 83
relation to colon–rectal cancer, 62
High-fiber foods
effect on fat absorption, 38
relation to cancer, 64
uses, 95
Hippocrates, 4
Hydrochloric acid, role in digestion, 48
Hydrocolloid bulking agents, 94
Hydrogenated vegetable fats, relation to cancer, 66

I

Ice cream, effect on serum sugar level, 71
Indigestion, causes, 4
Indoor exercise, examples, 43
Insoluble cereal fibers, cholesterol-reducing properties, 116
Insulin, 69,70
Intestinal bacteria, gas production, 57
Iranian males, zinc deficiency, 76
Iron
binding by hemicellulose, 77
binding by protein and sugars, 81
effect of pectins, 83
effect of phytic acid, 84
effect of wheat bran on absorption, 82
factors promoting absorption, 82
function, 76,93
mechanisms of binding, 81
release from bound state by enzymes, 80
Iron balance, effect of diet, 83
Irritable colon, effect of prolonged laxative use, 51

J

Journal of the National Cancer Institute, 65

K

Kay, R., 6
Kellogg's Crackiin Oat Bran, 108
Kellogg's Mueslix, 108
Kellogg's Nutrific Oatmeal, 104
Kelsay, J., 6
Kidney beans, effective binder of bile acids, 31
Kritchevsky, David, 6,28,29,66
Kwashiorkor, 91

L

Laxatives
cost to consumers, 48
effects of prolonged use, 50

Leavening process, effect on phytic acid, 84
Legumes
effect on serum sugar level, 71
health benefits for diabetics, 70
nutritional value, 119
source of phytic acid, 85
Lettuce
fiber content, 118
water-holding capacity, 55
Lifestyle
cancer risk factor, 66
effect on heart disease, 42
effect on obesity, 42
Lignin(s)
degradation during digestion, 56
dietary fiber, 8,16
interaction with bile acids, 31
cancer-preventing compound, 64
Lignin-containing fibers, effect on bile acid excretion, 29
Lima beans, fiber content, 118
Linoleic acid, relation to cancer, 66
Linseed, source of mucilage, 20
Lipase, role in digestion, 48
Lipids, 28
Lithocholic acid, 69
Liver, synthesis of cholesterol, 26
Living habits, relation to colon–rectal cancer, 63,64
Locust bean gum, 19
Loma Linda University, 67
Long-grain rice, binding of bile acids, 32
Louis Harris poll, results, 123
Low-density lipoprotein, bad cholesterol, 26
Low-fiber diets
effect on bowel function, 47
effect on mineral balance, 84
Lupin, pectin source, 17

M

Magnesium
binding by hemicellulose, 77
binding by phytic acid, 75,85
function, 76,93
Male sterility, effect of diet, 76
Malnutrition, 91
Manganese
binding by hemicellulose, 77
function, 93
Mannose, 15
Marasmus, 91
Medium-grain rice, binding of bile acids, 32
Methylcellulose, 13,94
Microcrystalline cellulose (Alphacel), use as food additive, 35
Miller, W. B., 13
Mineral(s)
binding by hemicellulose, 77
binding by pectins, 83–84
binding by wheat brans, 82–83
bioavailability factors, 77
bioavailability from high-fiber diets, 76
effect of minerals on absorption, 82
essentiality, 96
excretion in feces, 82
function, 76,93
interaction with fiber, 77
mechanisms of binding and release, 81
release from bound state by enzymes, 80
sources, 91
Mineral balance, effect of diet, 83
Mineral nutrition, relation to fiber, 75–86
Monomer, 11
Mucilages
degradation during digestion, 56
dietary fiber, 8
sources, 19, 20
types, 20
Mueslix Bran Cereal, 102
Murphy's Law, 97
Mustard, pectin source, 17
Mustard greens, anti-cancer food, 65

N

Nabisco, 101,103
Nabisco 100% Bran, 103
National Academy of Sciences
Committee on Diet, Nutrition, and Cancer, 112
Food and Nutrition Board, 112
National Cancer Institute, 93,97,101,112
National Food Consumption Survey, 40
Nationwide Food Consumption Survey, 65
Natural Bran Flakes, 103
Natural Raisin Bran, 101,103
Negative mineral balance, prevention, 83,86
Neutral detergent fiber (NDF)
in cereals, 103–104
method, 21
Neutral detergent residue, 94
New Orleans Dietetic Association, 43
New Orleans style red beans, recipe, 142
Noncaloric nutrients, 92
Nonpolysaccharide fiber, 16
Nutrient, definition, 95

Nutrient density, 89
Nutrient requirement, effect of aging, 62
Nutritional content of cereals, 102–103
Nutritional labeling of food products, 89
Nutritional problems associated with aging, 49–50
Nuts
  fiber content, 120–121
  oil content, 120

O

Oat(s)
  benefit to diabetics, 70,72
  effect on cholesterol levels, 29–30
  fiber content, 108
  high-grade feed, 108
Oat bran
  benefit to diabetics, 70
  lowering of blood cholesterol levels, 108,109
  mineral source, 75
  water-holding capacity, 55
Obesity
  control, 96
  effect of diet and lifestyle, 5,35,42
Okra, 118,120,136
Okra gumbo, 20
Oleic acid, health benefits, 28,92
Olive oil, 28,92
Olson, Alfred, 21
Omega–3-unsaturated fatty acids, health benefits, 92
Onions
  pectin source, 18
  water-holding capacity, 55
Orange
  fiber content, 115
  water-holding capacity, 55
Orange albedo, fiber content, 117
Orange pulp, water-holding capacity, 55
Oxalic acid, 77

P

Pancreatic lipase, effect of fiber, 37
Paper pulp, cellulose, 12
Parsnips, source of phytic acid, 85
Pasta, effect on serum sugar level, 71
Peach, fiber content, 115
Peanut(s)
  fiber content, 121
  oil content, 120
Peanut oil, 28,92
Peanut skins, fiber content, 121
Pear, fiber content, 115
Peas
  fiber content, 119
  nutritional value, 119
Pectins
  binding of heavy metals, 84
  binding of minerals, 83–84
  commercial sources, 17
  degradation, 56,58
  description, 16–18
  dietary fiber, 8
  effect on bile acid excretion, 29
  effect on copper balance, 83
  effect on serum cholesterol, 29
  water-holding capacity, 55
Pentosans, 8,15,82
Pepsin
  release of bound minerals, 80
  role in digestion, 48
Peptic ulcers, treatment with diet, 73
Phenolic acids, cancer-preventing compounds, 64
Phenylpropane molecules, lignin building block, 16
Phillips, Roland, 67
Phosphorus
  effect of pectins, 83
  function, 76,93
Physical activity, health benefit, 69
Phytate, *See* Phytic acid
Phytic acid
  binding of minerals, 75,80,85
  effect on fiber–mineral interactions, 84
  sources, 84,85
Phytin, *See* Phytic acid
Pioneers in dietary fiber research, 6
Plant extracts, source of gums, 19
Plant fibers, variation in bile acid-binding properties, 31
Plant foods
  examples, 11
  source of dietary fiber, 3
Plantix, 94
Plantix complex, 94
Plaques, fatty deposits, 28
Polymer, definition, 11
Polysaccharides, 11
Polyunsaturated fats, major sources, 27
Polyunsaturated fatty acid (PUFA), promotion of tumor growth, 66
Polyunsaturated vegetable oils, 28
Pork and beans, high-fiber food, 117–118
Post Oat Flakes, 108

Post Raisin Bran, 104
Potatoes
effect on serum sugar level, 71
effective binder of bile acids, 31
fiber content, 119
source of phytic acid, 85
water-holding capacity, 55
Powdered (food-grade) cellulose, 13
Precancerous polyps, effect of cereal fiber, 65
Prevention magazine, 123
Proper nutrition, 124
Propionic acid, mold-growth retardant, 115
Prostaglandins, 92
Prostate cancer, effect of animal protein, 68
Protein
binding of minerals, 80
caloric contribution to diet, 89
function, 80,91
malnutrition, 91
metabolism, 95
sources, 90
Prothrombin, 93
Prune, fiber content, 115
Prune juice, 116
Psyllium, health benefits, 19
Psyllium seed husks, source of mucilage, 19
Purified plant fiber, 94

Q

Quaker Corn Bran, 103,104
Quaker Oat Squares, 104

R

Raisin, fiber content, 114,115
Raisin Bran, 102
Raisin bran cereals, comparison, 102,104
Raisin Squares, 102
Ralston Bran Chex, 103,104,108
Ralston Oat Bran Options, 108
Rancidity, 93
Rapeseed oil, 28
Ready-to-eat breakfast products, 99
Ready-to-eat cereals, source of dietary fiber, 99
Recipes high in dietary fiber, 131–146
Recommended daily allowance (RDA) of nutrients, 90
Rectal gas, cause, 58
Rhamnose, 15,18
Rice
effect on serum sugar level, 71
fiber content, 119
neutral detergent fiber content, 106
Rice bran
characteristics, 107
fiber content, 106
hypoallergenicity, 107
mineral source, 75
stabilization, 107
water-holding capacity, 55
Rice bran hemicelluloses
effect on fecal water content, 52
effect on pancreatic lipase activity, 37
Rice cellulose, effect on lipid excretion, 36
Rice hemicelluloses
binding of bile acids, 32
description, 15–16
effect on cholesterol levels, 29–30
effect on lipid excretion, 36
Risk factors for colon-related cancer, 66
Root vegetables, fiber content, 119–120
Roughage, definition, 7,11
Ruminant stomachs, 12

S

Salad dressings
cottage cheese creamy dressing, 132
Dijon mustard–yogurt dressing, 132
vinaigrette dressing, 132–133
yogurt creamy dressing, 132
Salads
deluxe carrot salad, 133
deluxe cole slaw, 134
fresh summer vegetable salad, 136
fruit salad, 135
three-bean salad, 135
vegetable salad, 136
Waldorf salad without pecans or walnuts, 134
Salt content of breakfast cereals, 103
Salyers, Abigail, 57
Saturated fats,
effect on LDL (bad) cholesterol, 27
relation to cancer, 66
Saturday Evening Post, 89
Selenium, function, 93
Serum triglycerides, 25
Seventh Day Adventist Church, 67
Shortenings, relation to cancer, 66
Shredded Wheat 'N Bran, 101,103
Sigmoid colon, 48
Slippery elm bark, source of mucilage, 20
Snacks high in fiber, 143–144
Soluble fibers, cholesterol-reducing properties, 116

Soups
  bean soup, 139
  heart bean or lentil soup, 139
  hearty vegetable soup, 137–138
  lentil soup, 139
  vegetable soup, 137–138
Soups and gumbo, recipes, 136–139
Southgate, D., 6
Soybeans, pectin source, 18
Spiller, Gene, 6,94
Spinach
  fiber content, 119
  source of oxalic acid, 77
Squash, fiber content, 119
Starches
  effect on serum sugar level, 71
  properties, 12
  sources, 91
Steps to good health, 124–126
Strawberry, fiber content, 115
String beans, effective binder of bile acids, 31
Sugar(s), effect on pectin properties, 17
Sugar beet(s), pectin source, 17
Sugar beet pulp, water-holding capacity, 55
Sugar content of breakfast cereals, 102
Sugar metabolism, failure in diabetes, 70
Sugar polymers, 11
Sulfite treatment for storage of fruits, 116
Sunflower, pectin source, 17
Sunflower seed oil, 28
Sweet potato, fiber content, 119
Synthetic gum, 19

T

Taurocholic acid, binding by rice hemicelluloses, 32
Teenagers, food choices, 129–131
Ten-State Nutrition Survey, 65
Thickening agents, cellulose derivatives, 14
Thrombin, 93
Tooth loss, relation to fiber, 73
Total dietary fiber (TDF) method, 21
Toxic heavy metals, binding by pectins, 84
Tree exudates, source of gums, 19
Triglycerides, binding of bile acids, 30–33
Trowell, Hugh, 4,64
Trypsin
  release of bound minerals, 80
  role in digestion, 48
Turnip greens
  anti-cancer food, 65
  fiber content, 119

U

Unavailable carbohydrate, 94
Unbolted wheat meal, effect on bowels, 4
University of Georgia, 69
University of Illinois, 57
University of Lund, 69
Unmilled wheat flour, effect on bowels, 4
Unnatural fiber, 94
Unrefined cereals, relation to cancer, 64
Unsaturated fatty acids, promotion of tumor growth, 66
U.S. Department of Agriculture
  Southern Regional Research Center, 15
  Western Regional Research Center, 107
U.S. Food and Drug Administration, 13

V

Van Soest, P., 6
Varicose veins, effect of diet, 4,6,48
Vegetable(s)
  effect on iron balance, 83
  fiber content, 117
  health benefits due to fiber, 111–121
  types of dietary fiber, 112
Vegetable entrees, 140
Vegetable gums, soluble dietary fiber content, 20
Vegetable oils, 27
Vegetable trinity in soups, 137
Vegetarian lifestyle, health benefits, 68
Virginia Polytechnic Institute, 57
Vitamin(s)
  essentiality, 96
  functions, 92
  sources, 91
Vitamin A, function, 92
Vitamin C
  effect on binding of iron, 81
  function, 93
Vitamin D, function, 93
Vitamin E, function, 93
Vitamin K, function, 93

W

Walking, health benefits, 41
Walnuts, fiber content, 121
Water
  essentiality, 96
  function, 93
  role in body maintenance, 52

Water-holding properties of foods, 52,53,55
Weight, effect of diet, 5
Weight control
  factors, 35
  role of dieting, 38–40
  role of exercise, 41–44
  what not to do, 44–45
Western diets, characteristics, 65
Wheat bran
  advantage in cereals, 105
  binding of minerals, 81,82–83
  chemical properties, 82
  composition, 105
  effect on bile acid excretion, 29
  effect on mineral absorption, 82
  effect on mineral balance, 83
  effect on serum cholesterol, 29
  fiber content, 106
  interaction with bile acids, 31
  mineral source, 75
  vs. corn bran, 105
  water-holding capacity, 55
Wheat bran cereals
  effect on precancerous polyps, 65
  prevention of constipation, 104,109
Wheat flour, effect on bowels, 4
Wheat meal, effect on bowels, 4
Wheat pentosans, 82
Whole barley, fiber content, 106
Whole brown rice
  blood coagulation factor, 81
  effect on iron balance, 83
Whole cereal grains, source of phytic acid, 84
Whole-meal wheat bread, effect on iron balance, 83
Whole rice, fiber content, 106
Whole wheat, fiber content, 106
Whole wheat grains, advantage in cereals, 105
Wilson, Justin, 139
Wistar Institute, 28
Wood, source of food-grade cellulose, 13
Wood pulp
  cellulose, 12
  use as food additive, 35

X

Xanthan gum, 19
Xanthomonas campestris, source of gum, 19
Xylose, 15,82

Z

Zinc
  binding by hemicellulose, 77
  binding by phytic acid, 75,84,85
  binding by sugars, 81
  effect of pectins, 83
  effect of wheat bran on absorption, 82
  function, 76,93
  relation to male sterility, 76
  release during dough fermentation, 85
  release from bound state by enzymes, 80
Zinc deficiency, effect of diet, 76
Zucchini, fiber content, 119

Copy editing: Janet S. Dodd
Production: Peggy D. Smith
Indexing: A. Maureen Rouhi
Design and illustrations: Lisa Vann
Graphics support: Kathleen Schaner
Acquisition: Robin Giroux

Typeset by TypeWorks Plus, Inc., Silver Spring, MD
Printed and bound by Bookcrafters, Chelsea, MI

About the Author

Robert L. Ory received a B.S. in chemistry from Loyola University in New Orleans, an M.S. in organic chemistry from the University of Detroit, and a Ph.D. in biochemistry and nutrition from Texas A&M University. For almost 40 years, he has been active in research on oilseed proteins and enzymes; lipid oxidation mechanisms; the dietary fiber of cereals such as rice and barley; the enzymes involved in the quality of peanuts, rice, and several vegetables; and the interactions of rice dietary fiber with minerals and bile acids. In 1968 he was invited to start a research program on barley enzymes and proteins as a Fulbright—Hayes Research Scholar at the Polytechnic Institute of Denmark. In 1969 he returned to the Southern Regional Research Center in New Orleans to continue his research on rice and peanuts, where he remained until he retired in 1985. Since then he has been serving as Adjunct Professor of food science with the Virginia Polytechnic Institute and State University Extension Service.

He is a member of numerous professional societies, including the American Association of Cereal Chemists; the American Chemical Society (Division of Agricultural and Food Chemistry chairman in 1980, Distinguished Service award in 1986, named a Fellow in 1989); and the American Oil Chemists Society (chairman of the Protein and Co-Products Division in 1990, Award of Merit in 1987). He also has served as an associate editor on the boards of three professional journals.